ZEITSCHRIFT FÜR ANGEWANDTE GEOLOGIE

12

BAND 34

DEZEMBER 1988

ISSN 0044-2259
Z. angew. Geol., Berlin
34 (1988) 12, 353—384

Herausgeber:
Zentrales
Geologisches
Institut

AKADEMIE-VERLAG BERLIN

Zeitschrift für angewandte Geologie

BAND 34 | **1988** | **HEFT 12**

	INHALT	СОДЕРЖАНИЕ	CONTENTS
DVOŘÁK, J.; PAPROTH, E.	Trends der variszischen Entwicklung am SW-Rand der Osteuropäischen Tafel S. 353–359, 4 Abb., 44 Lit.	Тенденции варисцийского развития юго-западного борта Восточно-Европейской платформы стр. 353–359, 4 илл., 44 лит. ист.	Trends of the Variscan development near the SW border of the East European Platform pp. 353–359, 4 Figs., 44 Refs.
PATZER, U.; PRÖHL, S.	Was ist seismische Stratigraphie? Teil 2: Anwendungsmöglichkeiten S. 359–364, 7 Abb.	Что такое сейсмическая стратиграфия? Часть 2: возможности применения стр. 359–364, 7 илл.	What is seismic stratigraphy? Part 2: Possibilities of application pp. 359–364, 7 Figs.
MENZ, J.	Die Ableitung optimaler Erkundungsnetze auf der Grundlage einer geostatistischen Einschätzung der Verluste und Verdünnungen S. 364–368, 3 Abb., 5 Lit.	Выбор оптимальной разведочной сетки на основе геостатистической оценки потерь и разубоживания стр. 364–368, 3 илл., 5 лит. ист.	Deduction of optimum prospection nets on the base of geostatistical estimation of losts and dilutions pp. 364–368, 3 Figs., 5 Refs.
BACHMANN, H.	Ermittlung vorläufiger Konditionen bei der Gewinnung mineralischer Rohstoffe durch Laugung oder mikrobielle Prozesse S. 368–370, 3 Abb.	Определение временных кондиций при добыче минерального сырья посредством выщелачивания или с помощью процессов, связанных с деятельностью микроорганизмов стр. 368–370, 3 илл.	Investigation of preliminary conditions in mining mineral raw materials by leaching or bacteriological processes pp. 368–370, 3 Figs.
KAHNT, G.	Mathematisch-statistische Ableitung von Fehlerparametern für kohlechemische Analysendaten innerhalb der Erkundung fester Brennstoffe S. 371–374, 3 Tab., 6 Lit.	Математическо-статистическое определение параметров ошибок данных углехимических анализов в ходе разведки твердых горючих ископаемых стр. 371–374, 3 табл. 6 лит. ист.	Mathematic-statistical deduction of parameters of error for coal-chemical data within the prospection of solid fuels pp. 371–374, 3 Tables, 6 Refs.
SCHEUTZOW, R.	Quecksilber in Standardgesteinsproben und die Möglichkeit seiner Attestierung S. 374–376, 1 Abb., 3 Tab., 4 Lit.	Ртуть в стандартных образцах и возможности ее аттестации стр. 374–376, 1 илл., 3 табл., 4 лит. ист.	Mercury in standard rock samples and the possibility to its certification pp. 374–376, 1 Fig., 3 Tables, 4 Refs.
HEYNE, K.-H.; BOGNITZ, H.	Fehlerbetrachtungen zu markscheiderischen, boden- und gesteinsmechanischen Meßergebnissen und deren Auswirkung auf bergbausicherheitliche Parameter S. 376–382, 1 Abb., 9 Tab., 15 Lit.	Рассмотрение ошибочных результатов маркшейдерских измерений и данных по механическим свойствам грунтов и пород и их влияние на параметры безопасности стр. 376–382, 1 илл., 9 табл., 15 лит. ист.	Investigation into errors in surveying, soilmechanical and petrophysical measuring results and their effects on safety in mining pp. 376–382, 1 Fig., 9 Tables, 15 Refs.
Tagungsberichte	S. 382–383		
Buchbesprechungen	S. 383–384a		

Zeitschrift für angewandte Geologie is indexed in Current Contents / PC & ES.

Titelbild:
Farnsamer (*Palaeoweichselia de Francei* BRGT), Oberkarbon
Aufnahme: Dr. A. TÖLKE, Berlin

Die ZEITSCHRIFT FÜR ANGEWANDTE GEOLOGIE berichtet ständig über folgende Arbeitsgebiete: Geologische Grundlagenforschung und Lagerstättenforschung / Methodik der geologischen Erkundung / Ökonomie und Planung der geologischen Erkundung / Technik der geologischen Erkundung / Geologie und Lagerstättenkunde im Ausland. In der Zeitschrift können alle strittigen Fragen der praktischen Geologie behandelt werden. Die Autoren übernehmen für ihre Aufsätze die übliche Verantwortung.

Bezugsmöglichkeiten der „Zeitschrift für angewandte Geologie"
Bestellungen sind zu richten

- **in der DDR** an den Postzeitungsvertrieb unter Angabe der Kundennummer des Bestellenden oder an den Akademie-Verlag Berlin, Leipziger Straße 3–4, DDR-1086 Berlin;
- **im sozialistischen Ausland** an eine Buchhandlung für fremdsprachige Literatur oder an den zuständigen Postzeitungsvertrieb;
- **in der BRD und Berlin (West)** an eine Buchhandlung oder an die Auslieferungsstelle KUNST und WISSEN, Erich Bieber oHG, Postfach 102844, D-7000 Stuttgart 10;
- **in den übrigen westeuropäischen Ländern** an eine Buchhandlung oder an die Auslieferungsstelle KUNST und WISSEN, Erich Bieber GmbH, General Wille-Straße 4, CH-8002 Zürich;
- **im übrigen Ausland** an den Internationalen Buch- und Zeitschriftenhandel; den Buchexport, Volkseigener Außenhandelsbetrieb der Deutschen Demokratischen Republik, Postfach 160, DDR-7010 Leipzig, oder an den Akademie-Verlag Berlin, Leipziger Straße 3–4, DDR-1086 Berlin.

Zeitschrift für angewandte Geologie
Herausgeber: Zentrales Geologisches Institut, Berlin, im Auftrag des Ministeriums für Geologie der Deutschen Demokratischen Republik.
Chefredakteur: Doz. Dr. sc. B. KÖLBEL, Zentrales Geologisches Institut, Berlin, Fernruf: 2363997.
Anschrift der Redaktion: Kastanienallee 69, DDR-1054 Berlin; Fernruf: 2806334.
Verlag: Akademie-Verlag Berlin, Leipziger Straße 3–4, DDR-1086 Berlin; Fernruf: 2236201 oder 2236229; Telex-Nr.: 114420.
Bank: Staatsbank der DDR, Berlin Kto.-Nr.: 6836-26-20712.
Veröffentlicht unter der Lizenznummer 1279 des Presseamtes beim Vorsitzenden des Ministerrates der Deutschen Demokratischen Republik.
Gesamtherstellung: VEB Druckhaus „Maxim Gorki", DDR-7400 Altenburg.
Erscheinungsweise: Die Zeitschrift erscheint jährlich in einem Band zu 12 Heften. Bezugspreis je Band 156,– DM zuzüglich Versandspesen. Bezugspreis je Heft 13,– DM. Der gültige Jahresbezugspreis für die DDR ist der Postzeitungsliste zu entnehmen.
Bestellnummer dieses Heftes: 1047/34/12.
Alleinige Anzeigenannahme: DEWAG-Werbung, Wilhelm-Pieck-Straße 49, DDR-1054 Berlin, und alle DEWAG-Betriebe und -Zweigstellen der Bezirke der DDR.

AN (EDV) 46337
00220

ZEITSCHRIFT FÜR ANGEWANDTE GEOLOGIE

BAND 34 / 1988 / HEFT 12

Trends of the Variscan development near the SW border of the East European Platform

JAROSLAV DVOŘÁK, Brno; EVA PAPROTH, Krefeld

Introduction

From beginning to end, the development of the Variscides is marked by a continuous N/S pressure (DVOŘÁK 1985a). Sedimentation and deformation are manifestations of the same cause: the pressure of the Gondwana plate against the Laurasia plate (PEGRUM 1984), and cannot be separated one from the other. In central Europe, no undisputed facts are known that support a separation of north and south Europe and the existence of a Variscan ocean (BOUCOT & GRAY 1983). On the contrary, the numerous characters of facies and fossil distribution which are the most reliable in this respect contradict such hypothesis (JUBITZ, in print, lithological-palaeogeographical maps Givetian, Upper Viséan, Westphalian A/B). The Variscides are seen as a "give" zone (triple collision zone) on the granitic layer of the crust of Laurasia; the deformation certainly affected also parts of the mantle.

During the Variscan development, bordering parts of the Laurasia platform were pressed under the tectogene ("subfluence", WEBER 1978; DVOŘÁK 1978): Parts of them melted and ascended into internal parts of the tectogene where they formed granitoids. Consequently, the Variscan median mass (Zwischengebirge[1]) were uplifted and gave the clastic material into flysch-throughs. At the end graben-like intermontane basis formed.

The Mid-European Variscan tectogene is understood as a qualitatively symmetric construction (fig. 1*). On both flanks of the Alemannian-Moldanubian axis, appropriate belts may be correlated (KOSSMAT 1927; DVOŘÁK & PAPROTH 1969):

Variscan median mass:

- Proterozoic metamorphic structures: Mid-German Crystalline rise, Erzgebirge rise, Moldanubian, Moravian;
- Geosynklinal furrows (without polarity): Saxothuringian, Barrandian

Variscan outer geosynclines with pronounced polarity: Rhenohercynian, Sudetian

Variscan forelands: Brabant-Dutch-German platforms, Pre-Carpathian platform.

The bordering parts of the Proterozoic East European platform which formed the forelands of the tectogene have been distinguished as "Brabant" and "Dutch-German platforms" in the NW, as "Pre-Carpathian platform" in the SE. In the NE, parts of the Hamburg-Dobruja lineament (Odra lineament, DVOŘÁK 1973b) formed the boundary between the tectogene and a mobile part of the East European platform. Near the lineament, Variscan structures show SW vergency; this is particularly obvious in the vicinity of NW/SE directed faults, for instance in the Lausitzer Schiefergebirge and the Elbtal-Zone (GDR), in the western Sudetes (Poland), Krkonoše and Zlaté Hory ore structure in the Moravian Jeseníky mountains (Czechoslovakia).

In contrast to the compression structures in the tectogene and at the borders between tectogene and platforms, there are extension structures on the East European platform: the NW/SE directed Rügen-Lublin aulacogen sub-parallel to the Hamburg-Dobruja lineament (a compression structure), the Scottish SW/NE directed Midland Valley rift, and perhaps the NNW/SSE directed Ems rift east of Groringen (fig. 1).

The Tornquist line separated a more mobile part of the East European platform from the more stable part in the NE. The mobile part, particularly NE and SW of the Hamburg-Dobruja lineament, was broken into blocks and submitted to block faulting. NW/SE or NNW/SSE was one of the prefered fault directions on the northern foreland. Some of these "faults in the basement" can be tracket into the tectogene (fig. 1; DVOŘÁK 1973a, 1975).

The SE Pre-Carpathian platform was strongly influenced by the Alpidic tectogene, whereas the NW flank underwent only few important vertical and rotational movements since the end of Variscan time.

The development of the Mid-European Variscides is documented in the widespread Devonian and Carboniferous rocks. It seems difficult to determine exactly its beginning, because the oldest rocks are generally the most affected, so that older events are difficult or impossible to trace. Based on observations from the outer geosynclines in the NW (Rhenohercynian) and SE (Sudetian), the development may have started considerably earlier than the beginning of the Devonian (PAPROTH 1976). Pre-Devonian developments are often taken to refer to Caledonian movements. We prefer to look at them in this area as "early Variscan": beside the age there is no proof for Caledonian relations, and age is no admissible marker for a tectonic event. The numerous radiometric "Caledonian" ages found in the southern frame of the North Sea (ZIEGLER 1982) may proof temperature events of Cambrian-Silurian age in these areas; but these data are hardly influenced by Variscan thermal events too.

The end of the Variscan development had been reached near the Carboniferous-Permian boundary. The onset of a new

Eingang des Manuskripts in der Redaktion: 14. 7. 1987

[1]) Zwischengebirge. A term proposed by KOBER in 1921 to designate an undeformed or little deformed area between the two marginal chains of a symmetrical orogenic belt. In KOBER's original sense an orogen consists of two marginal chains and an intervening Zwischengebirge. The term has been translated into English as median mass (BUCHER 1933). — Glossary of Geology 1980.

* The tectonic subdivision given by the authors (fig. 1) is one of several versions discussed by the international working group of IGCP projekt no. 86. However, it does not agree with the version finally accepted for the Tectonic Map of the project. The project leadership

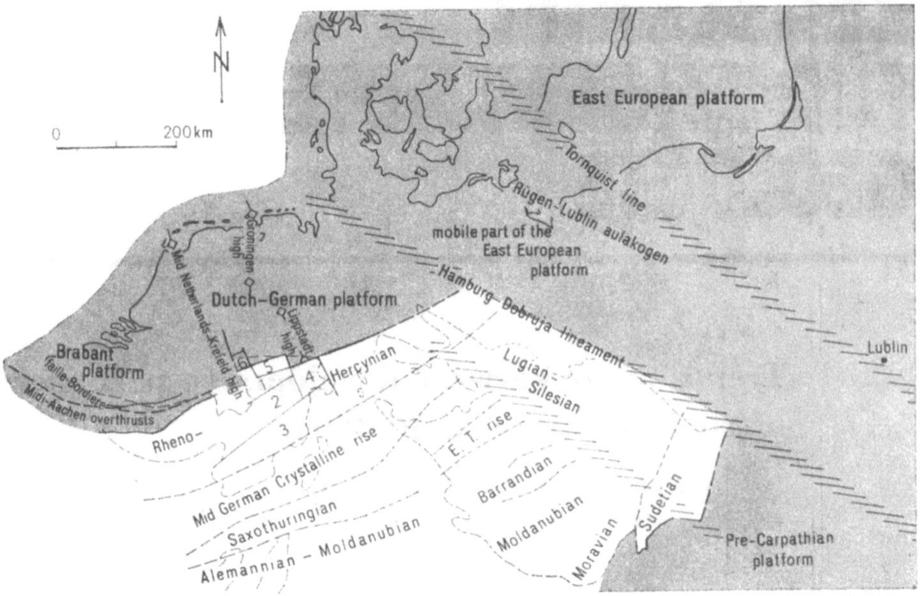

Fig. 1. Subdivision of the Mid-European Variscides and bordering platform areas
(mainly after Dvořák & Paproth 1969, Abb. 2; Paproth, Dresen & Thorez 1986, Fig. 4)
Platform areas in dark; E.-T. rise = Erzgebirge-Tepla rise
1 to 6 — inversion structures in the Rheinisches Schiefergebirge (Paproth 1976):
1 — Oberbergish-Sauerland inversion structure; 2 - Siegerland inversion structure; 3 — Nassau inversion structure; 4 — Ostsauerland inversion structure; 5 — Lenne inversion structure; 6 — Velbert inversion structure

dynamic, the Mesozoic, framework: the beginning of the plate tectonic Wilson cycle was marked by an intense short volcanic event (Rotliegend-Vulkanismus; Plein 1978). At the same time, the N/S directed Variscan pressure changed to an E/W directed extension demonstrated also by the Arctic-North Atlantic rift.

One of the most important — and characteristic — features of the Variscan tectogene was a high-temperature gradient during its development (Zwart 1976). In the Silesian rocks of the Variscan foredeeps, the contemporary temperature gradient was 70 °C/km in the NW (Westphalia; Buntebarth, Koppe & Teichmüller 1982) and 80 °C/km in the SE (Moravia; Dvořák, in print); in the median mass, the temperature gradient may locally have been even higher (200 °C/km).

Variscan trends in central Europe

In the development of the Mid-European Variscides, four stages may be distinguished (fig. 2). The dynamic framework was the same from the beginning of the sedimentation to the end of the tectonisation and consolidation. Depositional and deformational directions can therefore not be distinguished by their directions. As most of the movements shifted from the internal to the external belts (polarity), the beginning and the end of the stages may overlap considerably. Transgressional and regressional tendencies in the tectogene and its vicinity are contemporaneous and therefore supposed to be related to very widespread risings and droppings of sea level. It is not yet possible to clear this relationship.

Stage 1

Uplift of the forelands in the NW and SE of the tectogene. Subsidence of the geosynclines — development of inversion structures on the Rheinisches Schiefergebirge (Dvořák 1977) — which were gradually filled with thick siliciclastics in predominantly shallow and well turbulent water (Rhenish facies[2]). The siliciclastics were delivered mainly (80%) from the intensely weathered forelands, but also from parts of the Proterozoic median mass, e.g. the Mid-German Crystalline rise (Steemans 1986). On hillside of inversion structures, on top of the thick Rhenish facies sediments, thin beds of the Hercynian[2]) facies were deposited in remaining sedimentation areas, in shallow water. The rocks in the deeper parts of the inversion structures were folded and cleaved before the onset of the Stage 2. The synsedimentary development of the first axial cleavage (Roy 1978) is a proof of the development of the basin under compressional conditions (Dvořák 1973a).

Stage 2

Subsidence of the forelands: mid-Devonian transgression which was superimposed to the global Givetian transgressional trend. Widespread transgression on the forelands (e.g. southern North Sea area) and on the borders of the median mass too. Abundance of reef and reefoidal carbonates. The thickness of the carbonates depending on the rate of subsidence of the basement blocks may reach 1500 m (Balve-Brilon-Warstein areas, northern Rheinisches Schiefergebirge; Gotthardt, Mayer & Paproth 1978). Small amounts of siliciclastics were eroded from thoroughly weathered local sources into subsiding areas where Rhenish facies conditions outstayed. Near the borders of inversion structures, condensed sections of fine-grained sediments contain stratiform baryte and ore deposits (Meggen, on the northern border of the Siegerland structure — Dvořák 1987; Grube Eisen near the southern border of the Nassau structure).

In Moravia (Sudeticum) extensive marine transgression in the south- and east-directions during the Upper Eifelian and mainly Givetian was documented by numerous boreholes (Dvořák 1985b). A local transgression on the median mass is indicated in the Saar 1 borehole (Mid-German Crystalline rise) and by outcrops in the surroundings of Tišnov and Branná in Moravia (Moravian, Silesian).

In the Barrandian, in the axis area of the Variscan median mass, first flyschoid sediments appeared near the Eifelian-Givetian boundary: some hundred metres of siliciclastics mark here the end of the marine development.

A general subsidence and a paroxysm of spilitic volcanism (Stille's initial volcanism) supposedly marked a reorganisation in the deep crust and mantle in this time.

[2]) Rhenish facies — facies of well aerated, turbulent, shallow water. In contrast to Hercynian facies — facies of unsufficiently oxigenized quiet, shallow or deeper water (Schmidt 1931, pp. 20ff., 76ff.).

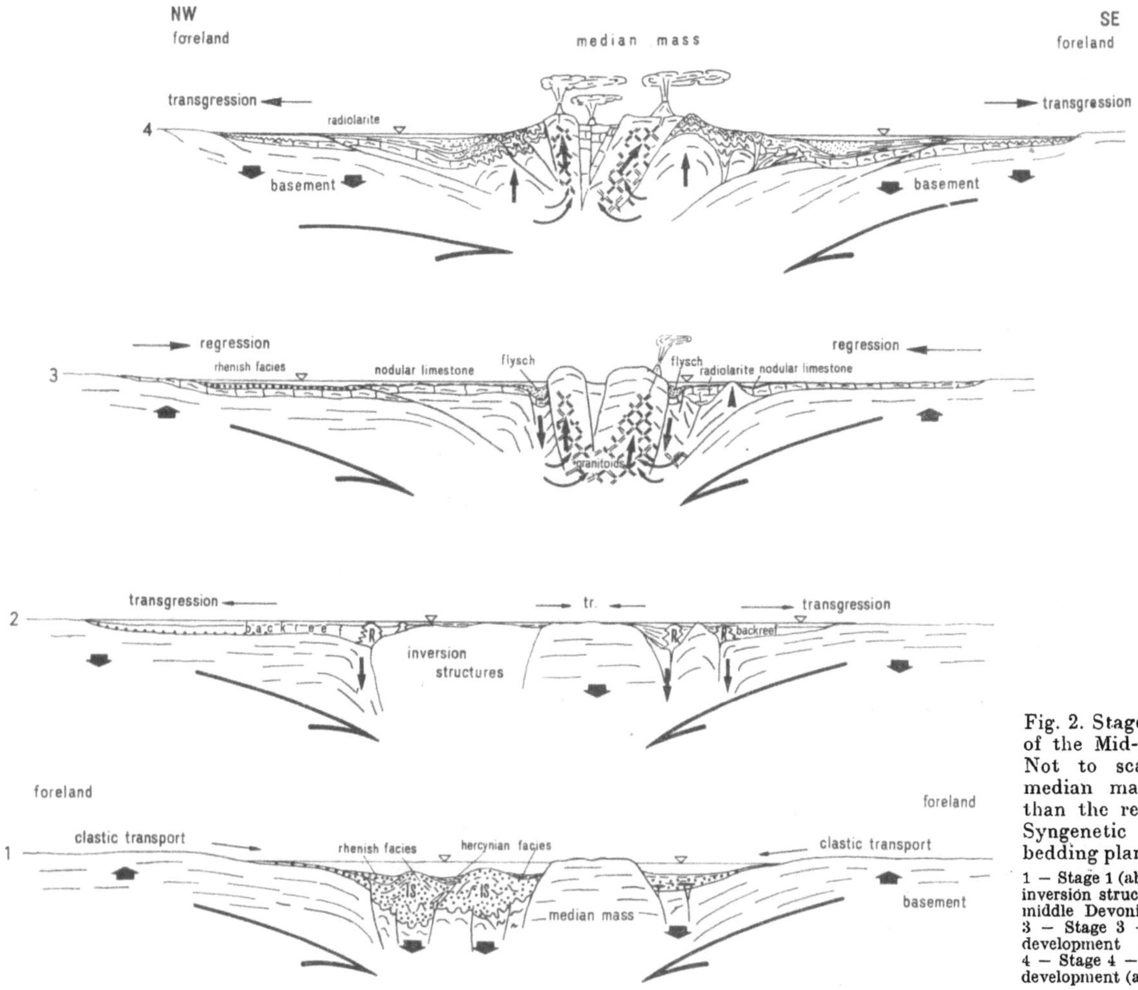

Fig. 2. Stages in the development of the Mid-European Variscides. Not to scale! In reality, the median mass is much broader than the rest of the Variscides! Syngenetic movements followed bedding planes

1 — Stage 1 (about early Devonian), IS — inversion structure; 2 — Stage 2 (about middle Devonian), tr. — transgression; 3 — Stage 3 — beginning of the flysch development (about late Devonian); 4 — Stage 4 — late stadium of the flysch development (about Viséan)

Stage 3

At the beginning of this stage slow lifting of the forelands started; Upper Frasnian—Famennian regression took place. The mosaic structure of the basement is reflected by the facies distribution on the forelands into the tectogene (Dvořák 1973a; Dvořák, Friáková & Lang 1976; Dreesen, Bless et al. 1985).

The quick lifting of many blocks of the median mass was related to ascending granitoid magmas, melted crust material that had been pressed from the forelands under the tectogene (subfluence). Acid volcanoes topped the mountains of the median mass. These upper parts were thoroughly eroded, as rhyolite fragments and volcanic quartz fragments are abundant in the greywackes (sic!), and the first biotite-bearing tuffs and tuffitic layers are well known from the flysch sediments and from the limestones of the neighbouring forelands (Dvořák 1985b). Volcanism marked the intrusion of granitoid magmas in deeper parts of the median mass. The large amounts of granitoids on the surface mark the places of the late Devonian and Carboniferous acid volcanism.

In the geosynclinal basin, in proximal parts (near the median mass) first flysch sediments (greywackes) were deposited: the fresh material came from quickly rising source areas in the median mass (Schmidt 1931; Engel, Franke et al. 1983 for the Rhenohercynian; Dvořák 1978 for the Sudetian). The belts of maximum subsidence were narrow, their axes ran subparallel to the axes of the rising source areas in the median mass. The transport directions of the siliciclastics into the basin were perpendicular to the axes, whereas the transport in the basins by sea currents and turbidity currents were parallel to the basin axis.

The basin axes shifted in the direction from the median mass to the forelands and caused proximal parts of the depositional areas to incorporate in the rising median mass. They show therefore regressive tendencies, whereas the distal parts — in the direction of the forelands — show transgressive tendencies. The speed of this shifting accelerated towards the end of the Stage 3, more clastic material was delivered into the migrating basins in a certain time, and thicknesses of individual formations and grain sizes increased.

The incorporation of the flysch deposits in the tectogene (s.s.) means that they were folded and metamorphosed together with metamorphic rocks of Proterozoic age, that they were eroded after their uplift and delivered siliciclastics into younger basins.

Outside the flysch areas, starved basins with reduced thicknesses of dark coloured shales and cherts developed near the areas with spilitic volcanism. On the SE flank of the tectogene, in Moravia the transgression on the foreland that was caused by the foreland-wards shifting of the basin axis is well known from boreholes (Dvořák 1985b): Frasnian reefoid limestones are overlain by late Viséan (V3b) sands that change upwards into biodetrital coral-brachiopod limestones (V3b—V3c).

The development of inversion structures continued in the Rhenohercynian (inversion structures 4 to 6). Elevated parts of the inversion structures were eroded or covered by thin, shallow water nodular limestones or red shales (Dvořák 1985b).

Stage 3 is characterized by increased compression of the tectogene. The regressional tendency due to lifting of the median mass and the tectogene's framework (Famennian regression on the forelands) was superimposed to regressional tendencies at the edges of the contemporaneous Old Red continent that may have been related to a global (?) dropping of sea level (BLESS & STREEL 1986).

Stage 4

The molasse stage was reached when sedimentation caught up with subsidence (DVOŘÁK 1975). The construction of the whole tectogene was mainly finished; but the tectogene and its forelands were still considerably uplifted, as the contemporaneous global Moscovian (about Westphalian) transgression did not reach the tectogene nor its forelands (BERGSTRÖM, BLESS & PAPROTH 1985), (fig. 3).
An early molasse is known from the Drahany upland (Sudetian, Moravia). Youngest sediments in the Variscan cycle are about 3000 m thick coarse conglomerates of late Viséan age. Rounded granulite blocks reaching 2 m in diameter were transported from a minimum distance of 50 km. Laterally, the conglomerate interfingers with greywackes and varve-like siltstones and shales. Most of the material was delivered from the katazonal metamorphosed Moldanubian of the Bohemian-Moravian high.
Coal formed over wide areas of the foredeeps and the forelands. Alternation of coal seams and siliciclastics is regular (cyclic) in the foredeeps and on the forelands, irregular in the intermontane basins. The often extremely pure and several metres thick coal seams mark long times practically without siliciclastic transport in large areas, e.g. the North-European paralic coal basin. In the paralic ("near to the sea") areas, deposition of siliciclastics generally started with a "drowning" of the coal forming vegetation in marine or non-marine waters. After the deposition of few metres thick fine-grained dark sediments, several metres thick sandy material was distributed by fluviatile transport and deposited. The sandy material became finer grained upwards and rootlet beds indicate the reappearance of the vegetation that generally led to a renewal of coal formation. Coal formation evidently lasted considerably longer than the short-time deposition of siliciclastics. Cyclothems and coal seams are not exactly isochronous deposits, as tonsteins cut them (e.g. Karl tonstein, WOLF 1971).

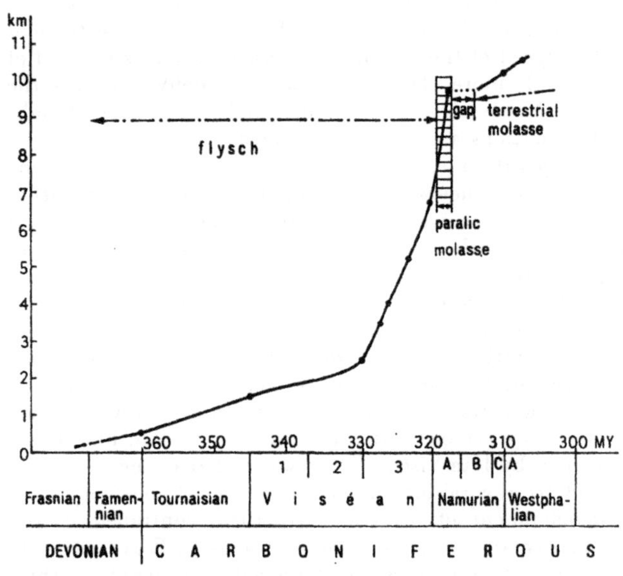

Fig. 3. Thickness of flysch and molasse sediments versus time within the zones of maximum subsidence migrating from the median mass to the foreland (Sudetian, Moravia, ČSSR)

Grabenlike intermontane basins were filled with several thousand metres of siliciclastics with intercalated coal seams. In general, coal seams may be thicker in intermontane basins than in paralic basins, but they are rather irregular. Coals are concentrated in inner parts of the basins, whereas coarse-grained siliciclastics predominate near the borders.
The large masses of detritus distributed on the forelands and in the intermontane basins were eroded from the rising Variscan mountains. The thickness of the sequence, the alternation with coal formation, the scarcety of conglomerates, and the fresh (unweathered) detritus indicate uniform and intense erosion, faster than weathering that was faciliated by a tropical climate. The extremely regular and uniform subsidence of the large foredeeps and forelands (coal-bearing cyclothems) reflects important tectonic processes. The same is true for the intermontane basins which seem to have been filled in correspondence to their intermittent subsidence.
The continuous uplift of the whole area — tectogene and forelands — is indicated by the general decrease of marine influence in the paralic forelands, from the Namurian on, which practically stopped near the Westphalian B—C boundary in NW. This is the more remarkable as the Moscovian transgression — one of the widest-spread in geological history — conquered large parts of the globe at the same time.
The development of the Upper Silesian coal field in Moravia and Poland is remarkable (fig. 4), because it started as a paralic coal-bearing foredeep and continued in NE parts as an intermontane one: the paralic part of the coal measures started in the earliest Namurian A (about the boundary E1a/E1b) and filled the typical foredeep. The about 3000 m thick beds are of early and middle Namurian A age (Pendleian

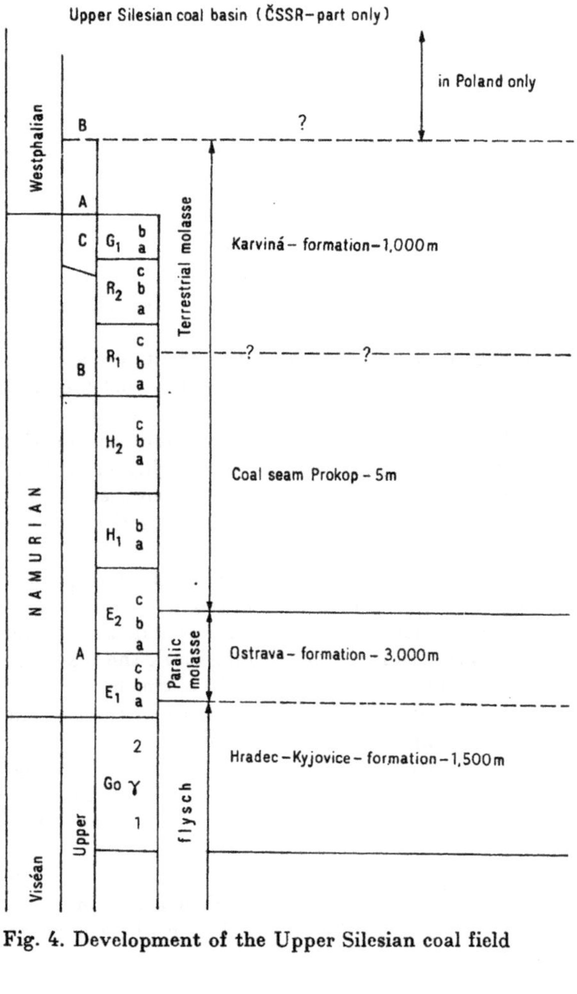

Fig. 4. Development of the Upper Silesian coal field

and Arnsbergian). Tuff layers are common (the 10 m thick Whetstone bank). The thickest coal seam is the youngest, situated on the boundary between the paralic and terrestrial sequence, coal seam Prokop with 5 m. The deposition of the material for the Prokop seam happened in a long crucial time: Floral elements in the coal seam are of Namurian A to B age, above the coal seam they indicate late Namurian B age. — Prevailing sediment colors change from dark grey in the paralic part to light colors in the continental part. The time of the coal seam Prokop deposition was the end of the main tectogenesis (folding) in Sudetian (Moravia).

In NW Germany, the formation of the Subvariscan foredeep started in the earliest Namurian. This basin was filled from source areas in the rising median mass with more than 2000 m thick Namurian A and B marine barren measures (Flözleeres). The coal-bearing sequence started here near the base of the Namurian C, later than elsewhere because of the greater mobility of the basement (i.e. absence of the pre-Variscan granitoids in the basement).

Intermontane basins

The oldest intermontane basins are of late Viséan age. They increase in number, surface area, and thickness of siliciclastic deposits during the Silesian. From the Westphalian C on, sediment thicknesses are greater in the intermontane basins than in the forelands. In Stephanian times only few hundred metres of rocks were deposited in the forelands and in relatively small areas, against several thousand metres in intermontane basins (Bless, Bouckaert et al. 1977). Coal-bearing facies continued into Stephanian and early Permian, when coal formation had stopped in the forelands.

The intermontane basins are graben-like features paralleling Variscan or Proterozoic structures. In the core of the Bohemian massif NW/SE and NE/SW (later NNE/SSW) directions prevail. These dilatational structures developed in the highest parts of the uplifted median mass in times of maximum compression of the whole tectogene.

Coalification

The relation between coalification rank of certain beds and their position in the tectogene or on the platform is complex. Local and regional post-Variscan thermal processes complicate the relation particularly in the forelands.

Measurements of local and regional differences in coalification rank may permit the reconstruction of palaeoheat flow and, thus, faciliate to trace tectogenetic processes. This has been studied particularly in the Rhenohercynian (Wolf 1972, 1978; Dvořák & Skoček 1975; Paproth & Wolf 1973; Paproth 1976; Lütke & Koch 1983), the Sudetian (Dvořák & Skoček 1975; Dvořák & Wolf 1979) and bordering foredeep and foreland areas in NW Germany (Teichmüller, Teichmüller & Bartenstein 1979; Teichmüller & Teichmüller 1986) and upper Silesia (Kotas 1971; Dvořák, in print).

In these areas low metamorphosis permits the observation of the coalification rank. In the Upper Silesian basin and Ruhr basin (in the foredeeps and on the borders of the platforms), coalification was finished before the beginning of the main folding (Teichmüller & Teichmüller 1971; Paproth & Wolf 1973), but in the tectogene (s.s.) the peak of the thermal metamorphosis was reached later — after the folding (the 2400 m deep Dětřichov 1 borehole in Moravia, the 1463 m deep Wulpke 2 borehole in Westharz). Generally spoken, maximum coalification is mainly determined by the temperature, if considerable differences in length of time of the heating may be excluded. The relative uniformity of coalification rank of sediments in the bordering forelands and in the tectogene points to a common heat source. Teichmüller & Teichmüller (1986) assumed that the heat source is caused by the reduced thickness of the crust below the foredeep of the tectogene, but the Paleozoic sediments in the Sudetian are more metamorphosed than the sediments of the Rhenohercynian. From this point of view the heating centres were extensive granitoids in the median mass.

Conclusion

The Variscan tectogene developed in a crustal triple collision zone of the Proterozoic Laurasian plate as it collided with the Gondwana plate. The "give" zone is characterized by a reduced thickness of the crust which may explain the high mobility of the Variscan area. The NW border of the East European platform approximated the trend of the Laurasian "give" zone: subdivided near the tectogene into Brabant and Dutch-German platforms in the NW, mobile parts of the East European platform in the NE and the Pre-Carpathian platform in the SE. The southern parts of the Variscan tectogene and its SE foreland were heavily affected — partly beyond recognition — by Alpidic movements.

In the course of Variscan tectogenesis, parts of the forelands evaded the lasting N/S (NW/SE) pressure by diving under the tectogene (subfluence). The high temperature gradient caused the melting of the material which ascended as granitoid magmas in inner parts of the tectogene. On the surface, the intrusions caused volcanism which produced the abundant acidic tuff and tuffitic layers known particularly from the uppermost Devonian and Carboniferous flysch and coal-bearing molasse deposits.

The behaviour of the forelands strongly influenced the development of the tectogene. In a first stage — in the early Devonian — the developing basins of the tectogene were filled mainly with siliciclastic material from the deeply weathered Proterozoic crystalline rocks (forelands) and, due to the regressional tendency during that stage, from parts of the Proterozoic median mass.

A second stage is marked by the predominance of carbonate sediments in the geosynclinal basins and in the forelands. This demonstrates not alone a widespread marine transgressional tendency which is related to the global Givetian transgression, but also an epoch in the development of the Variscan tectogen in these parts of the world: it marked a turning point before the beginning of the acute stages of the tectogen when the median mass raised continuously (in relation to the intrusion of granitoid magmas), until the end of the Variscan development at the end of the Carboniferous.

First flysch sediments in the Variscan geosynclines characterize a 3rd stage that is marked in the forelands by the Upper Frasnian and Famennian regressional trend. Siliciclastics were delivered into the geosynclinal basins from the rising median mass (greywacke), although well sorted sands spread on the fringes of the continental foreland areas (Old Red continent).

This substage was followed by the acme of flysch sedimentation. Increasing relief energy in the rising median mass is demonstrated by the bulk of the flysch greywackes. Sedimentation could not catch up with subsidence.

A 4th stage, the molasse stage, was reached when sedimentation caught up with subsidence. Coal measures developed on consolidated grounds, in the intermontane basins of the tectogene, and in the forelands. The largest of the latter is the central European paralic coal basin, spreading from Ireland/Scotland to central east Poland, in the Westphalian.

Near the end of the Variscan development, sedimentation decreased on the forelands, whereas it continued in growing intermontane basins, demonstrating lasting subsidence in the tectogene. It was only near the Carboniferous-Permian boundary that a fundamental change of the dynamic framework marked the beginning of a plate-tectonic Wilson cycle. An extensive volcanism marked this time in the Variscides; some thousand metres of volcanic rocks occurred in the north German-Polish lowlands (the mobile part of the East European platform without Variscan consolidation).

"*Despite various attempts, modern concepts of plate tectonics do not seem to be able to explain the Variscan orogeny which first of all is a widespread thermal event.*" (ZWART 1976; p. 367).

Summary

The Mid-European Variscides developed in the granitic layer of the crust, in a "give" zone (triple collision zone) of the Laurasian plate when it collided with the Gondwana plate. A lasting NW/SE pressure determined the sedimentary and tectonic development. A thin crust in the "give" zone may explain the high mobility of the basement. Subfluence of bordering mobile parts of the East European platform, the melting and subsequent ascendance of granitoid magmas in inner parts of the tectogene mark the high-temperature gradient.

Four stages may be distinguished in the development: 1) Filling of geosynclinal basins from source areas on the forelands mainly (about early Devonian). 2) Decrease of siliciclastic deposition, predominance of carbonate sediments during a general transgressional trend; reorganisation of the crust and mantle (about middle Devonian). 3) Uplift of the median mass in relation to intrusion of granitoid magmas, regression on the forelands (about late Devonian). Predominance of flysch deposits — formation of zones of maximal subsidence in the tectogene; subsidence could not be equilibrated by the sedimentation; first intermontane basins (about early Carboniferous). 4) Predominance of molasse deposits: subsidence was equilibrated by sedimentation; coal formation (about late Carboniferous).

The development of the Variscides terminated in central Europe near the Carboniferous-Permian boundary, when the beginning of plate tectonic Wilson cycle was marked by a volcanic paroxysm (Rotliegendvolcanism).

Zusammenfassung

Die mitteleuropäischen Varisziden haben sich in einer „Knautschzone" der Laurasiatischen Platte entwickelt, als diese mit der Gondwana-Platte kollidierte. Sedimentation und Tektonik waren von andauerndem NW—SE gerichtetem Druck bestimmt. Die dünne Kruste in der „Knautschzone" kann die hohe Mobilität erklären, die wiederum das Schmelzen randlicher Teile der Osteuropäischen Plattform, die durch Subfluenz unter das Tektogen gepreßt worden waren, ermöglichte, sowie ihren Aufstieg als granitoide Magmen in das Zwischengebirge.

Vier Entwicklungsstadien werden unterschieden: 1. Geosynklinalbecken werden hauptsächlich von den Vorländern mit Siliziklastika gefüllt (etwa frühes Devon). 2. Karbonatische Sedimente verdrängen die Siliziklastika zur Zeit einer allgemeinen marinen Transgression; Reorganisation der Kruste und des Mantels (etwa mittleres Devon). 3. Hebung des Zwischengebirges in Verbindung mit Intrusion granitoider Magmen; Regression auf den Vorländern (etwa spätes Devon) — Ausbildung von Zonen maximaler Absenkung im Tektogen; Vorherrschen von Flyschsedimenten: die Sedimentation konnte die Subsidenz nicht ausgleichen; erste intramontane Becken (etwa frühes Karbon). 4. Vorherrschen von Molassesedimenten: Subsidenz wurde durch die Sedimentation ausgeglichen; Kohlebildung (etwa spätes Karbon).

Die Entwicklung des Variszikums endete in Mitteleuropa etwa an der Karbon-Perm-Grenze. Der Beginn eines plattentektonischen Wilson-Zyklus wurde durch einen Höhepunkt in der vulkanischen Tätigkeit angezeigt (Rotliegend-Vulkanismus).

Резюме

Средне-Европейские варисциды сформировались в условиях „смятой зоны" Лавроазиатской плиты в период ее столкновения с плитой Гондваны. Осадконакопление и тектоника определялись длительными напряжениями, направленными с северо-запада на юго-восток. Наличием тонкой коры в „смятой зоне" можно объяснить её высокую мобильность, которая в свою очередь объясняет расплавление краевых частей Восточно-Европейской платформы, которые были затянуты под тектоген. С этим же фактом можно связать их последующий подъём в форме гранитоидных магм в межгорные поднятия. Выделяются 4 стадии развития:

1. геосинклинальные бассейны заполняются главным образом силицитовой кластикой, привносимой из областей предгорий (приблизительно ранний девон); 2. в период общей морской трансгрессии силицитовая кластика вытесняется карбонатными образованиями, происходит реорганизация коры и мантии (приблизительно средний девон); 3. воздымание межгорных поднятий в связи с интрузией гранитоидных магм; регрессия в областях форланда (приблизительно поздний девон); формирование зон максимального погружения в тектогене, преобладание флишевого осадконакопления, которое не смогло компенсировать погружение; первые межгорные бассейны (приблизительно ранний крабон); 4. преобладание молассовых образований: погружение компенсировалось осадконакоплением; угленакопление (приблизительно поздний карбон).

Развитие варисцид закончилось в Средней Европе где-то на границе карбона и перми. Начало плитно-тектонического цикла по Уилсону было отмечено максимумом вулканической деятельности (вулканизм в ротлигендесе).

References

BERGSTRÖM, J.; BLESS, M. J. M.; PAPROTH, E.: The Marine Knabberud-Limestone in the Oslo-Graben: Possible Implications for the Model of the Silesian Palaeogeography. — Z. dt. geol. Gesellsch., Hannover **136** (1985) S. 181—194

BLESS, M. J. M.; BOUCKAERT, J.; CALVER, M. A.; GRAULICH, J. M.; PAPROTH, E.: Palaeogeography of Upper Westphalian deposits in NW Europe with reference to the Westphalian C North of the mobile Variscan belt. — Mededel. Rijks geol. Dienst, n. Ser., Heerlen **28** (1977) S. 101—147

BLESS, M. J. M.; STREEL, M. (Eds.): Late Devonian events around the Old Red Continent. — Ann. Soc. géol. Belg., Liège **109** (1986) S. 1—323

BOUCOT, A. J.; GRAY, J.: A Paleozoic Pangaea. — Science, Washington **222** (1983) S. 571—581

BUNTEBARTH, B.; KOPPE, I.; TEICHMÜLLER, M.: Palaeogeothermics in the Ruhr basin. — In: ČERMÁK, V.; HAENEL, R. (Eds.): Geothermics and geothermal Energy. Stuttgart: E. Schweizerbart'sche Verlagsbuchhandlung 1982. S. 45—55

DREESEN, R.: A propos des niveaux d'oolithes ferrugineuses de l'Ardenne et du volcanisme synsédimentaire dans le Massif Ardenno-Rhénan au Dévonien Supérieur. Essai de corrélation stratigraphique. — Neues Jb. Geol. u. Paläont. Mh., Stuttgart (1982) S. 1—11

DREESEN, R.; BLESS, M. J. M.; CONIL, R.; FLAJS, G.; LASCHET, C.: Depositional environment, paleoecology and diagenetic history of the „Marbre rouge à crinoides de Baelen" (Late Upper Devonian, Verviers synclinorium, eastern Belgium). — Ann. Soc. géol. Belg., Liège **108** (1985) S. 311—359

DVOŘÁK, J.: Geologie des oberen Teiles der Myslejovice-Schichtenfolge am östlichen Rande der Drahaner Höhe. — Čas., morav. Muz. v Brnč, Vědy prir., Brno **54** (1969) S. 45—60

— Synsedimentary tectonics of the paleozoic of the Drahany Upland (Sudeticum, Moravia, Czechoslovakia). — Tectonophysics, Amsterdam **17** (1973) S. 359—391 (1973a)

— Problem concerning the northeastern closure of the Variscan orogene. — Neues Jb. Geol. u. Paläont. Mh., Stuttgart (1973) S. 449—454 (1973b)

— Interrelationship between the sedimentation rate and subsidence during the flysch and molasse stage of the Variscan geosyncline in Moravia (Sudeticum). — Neues Jb. Geol. u. Paläont. Mh., Stuttgart (1975) S. 339—342

— Inversion structures of the European Variscan orogene. — Včst. ústř. Úst. geol., Praha **52** (1977) S. 55—58

— Proterozoischer Untergrund der variscischen Geosynclinale in Mähren (ČSSR) und ihre Entwicklung. — Z. dt. geol. Gesellsch., Hannover **129** (1978) S. 383—390

— The red shales of the Upper Devonian in the Rhenish Slate Mountains. — Neues Jb. Geol. u. Paläont. Mh., Stuttgart (1985) S. 329—339 (1985a)

— Tektogeneze devonu a spodního karbonu Hornoslezské pánve. — Zemni plyn a nafta, Hodonín **30** (1985) S. 1—14 (1985b)

— Geological conditions of the genesis of the Meggen-type Deposits. — Věst. ústř Úst. geol., Praha **62** (1987) S. 77—82

— Geology of the Paleozoic sediments of the deep borehole Jablůnka 1 (Beskydy Mts. NE Moravia) — comparison with the deep borehole Münsterland 1. — Sb. geol. Věd., Praha (in print)

DVOŘÁK, J.; FRIÁKOVÁ, O.; LANG, L.: Block structure of the old basement as indicated by the facies development of the Devonian and the Carboniferous in the Moravian Karst (Sudeticum, Moravia, ČSSR). — Geol. et Palaeont., Marburg **10** (1976) S. 153—160

DVOŘÁK, J.; PAPROTH, E.: Über die Position und die Tektogenese des Rhenohercynikums und des Sudetikums in den mitteleuropäischen Varisziden. — Neues Jb. Geol. u. Paläont. Mh., Stuttgart (1969) S. 65—88

DVOŘÁK, J.; SKOČEK, V.: Reconstruction of the paleoheat flow regime in two areas in the Variscan Orogene. — Neues Jb. Geol. u. Paläont. Mh., Stuttgart (1975) S. 517—527

DVOŘÁK, J.; WOLF, M.: Thermal metamorphism in the Moravian Palaeozoic (Sudeticum, ČSSR). - Neues Jb. Geol. u. Paläont. Mh., Stuttgart (1979) S. 596 to 607

ENGEL, W.; FRANKE, W.; GROTE, C.; WEBER, K.; AHRENDT, H.; EDER, F. W.: Nappe Tectonics in the Southeastern Part of the Rheinisches Schiefergebirge. — In: MARTIN, H.; EDER, F. W. (Eds.): Intracontinental Fold Belts. Heidelberg: Springer-Verlag 1983. S. 267—287

GOTTHARDT, R.; MAYER, O.; PAPROTH, E.: Gibt es Massenkalke im tiefen Untergrund NW-Deutschlands, und können sie Kohlenwasserstoffe führen? — Neues Jb. Geol. u. Paläont. Mh., Stuttgart (1978) S. 13—24

KOSSMAT, F.: Gliederung des varistischen Gebirgsbaues. — Abh. sächs. geol. Landesamt, Leipzig **1** (1927) S. 1—39

KOTAS, A.: Uwagi o metamorfizme węgla Zagłebia Górno-Śląskiego. – Zesz. nauk. Acad. gór.-hutn., Ser. Geol., Kraków **292** (1971) 14, S. 7–25
LÜTKE, F.; KOCH, J.: Das Inkohlungsbild des Paläozoikums im Westharz und seine Interpretation. – Geol. Jb., R. A, Hannover **69** (1983) S. 3–42
PAPROTH, E.: Zur Folge und Entwicklung der Tröge und Vortiefen im Gebiet des Rheinischen Schiefergebirges und seiner Vorländer, vom Gedinne (Unterdevon) bis zum Namur (Silesium). – Nova Acta Leopoldina N. F., Leipzig **224** (1976) 45, S. 45–58
PAPROTH, E.; DREESEN, R.; THOREZ, J.: Famennian paleogeography and event stratigraphy of northwestern Europe. – Ann. Soc. géol. Belg., Liège **109** (1986) S. 175–186
PAPROTH, E.; WOLF, M.: Zur paläogeographischen Deutung der Inkohlung im Devon und Karbon des nördlichen Rheinischen Schiefergebirges. – Neues Jb. Geol. u. Paläont. Mh., Stuttgart (1973) S. 469–493
PEGRUM, R. M.: Structural development of the southwestern margin of the Russian-Fennoscandian Platform. – In: Petroleum Geol. North European Margin. London: Graham and Trotman 1984. S. 359–369
PLEIN, E.: Rotliegend-Ablagerungen im Norddeutschen Becken. – Z. dt. geol. Gesellsch., Hannover **129** (1978) S. 71–97
SCHMIDT, H.: Die ursprünglichen Zusammenhänge zwischen Harz und Rheinischem Schiefergebirge. – Naturwissenschaften, Berlin **19** (1931) S. 911–916
– Die bionomische Einteilung der fossilen Meeresböden. – Fortschr. Geol. u. Paläont., Berlin **12** (1935) S. 1–154
STEEMANS, P.: Palynostratigraphie de l'Eodevonien. – Thèse, Doc. Sci., groupe Sci. géol. mineral., Univ. Liège, Fac. Sci., Liège 1986. 358 S.
STILLE, H.: Über Alter und Art der Phasen variscischer Gebirgsbildung. – Nachr. K. Gesellsch. Wiss. Göttingen, math.-phys. Kl., Göttingen (1920) S. 218–224
TEICHMÜLLER, M.; TEICHMÜLLER, R.: Inkohlung. – Fortschr. Geol. Rheinland u. Westfalen, Krefeld **19** (1971) S. 47–56
–• Relations between coalification and palaeogeothermics in Variscan and Alpidic foredeeps of western Europe. – Lecture Notes in Earth Sci., Berlin(West), Heidelberg **5** (1986) S. 53–78
TEICHMÜLLER, M.; TEICHMÜLLER, R.; BARTENSTEIN, H.: Inkohlung und Erdgas in Nordwestdeutschland. Eine Inkohlungskarte der Oberfläche des Oberkarbons. – Fortschr. Geol. Rheinland u. Westfalen, Krefeld **27** (1979) S. 137–170
WEBER, K.: Das Bewegungsbild im Rhenoherzynikum – Abbild einer varistischen Subfluenz. – Z. dt. geol. Gesellsch., Hannover **129** (1978) S. 249–281
WOLF, M.: Vorkommen und Ausbildung des Kaolin-Kohlentonsteins in den Karl-Flözen Westfal A. – 6ᵉ Congr. Internat. Stratigr. Géol. Carbonif., Sheffield 1967. C. R., Maastricht **4** (1971) S. 1593–1604
– Beziehungen zwischen Inkohlung und Geotektonik im nördlichen Rheinischen Schiefergebirge. – Neues Jb. Geol. u. Paläont. Abh., Stuttgart **141** (1972) S. 222–257
– Inkohlungsuntersuchungen im Hunsrück (Rheinisches Schiefergebirge). – Z. dt. geol. Gesellsch., Hannover **129** (1978) S. 217–227
ZIEGLER, P. A.: Geological Atlas of Western and Central Europe. – s'Gravenhage: Shell Internationale Petroleum Maatsch. B. V. 1982
ZWART, H. J.: Regional metamorphism in the Variscan orogeny of Europe. – Nova Acta Leopoldina, N. F., Leipzig **224** (1976) S. 361–367

Was ist seismische Stratigraphie?

Teil 2: Anwendungsmöglichkeiten

ULRICH PATZER, STEPHAN PRÖHL, Leipzig

Mitteilung aus dem VEB Kombinat Geophysik Leipzig

Nachdem im ersten Teil dieser Arbeit[1]) wesentliche Begriffe und die methodischen Grundlagen der seismischen Stratigraphie erläutert worden sind, sollen im zweiten und abschließenden Teil die Aspekte der praktischen Nutzung dieses Konzeptes diskutiert werden.

Den bereits im vorigen Teil genannten prinzipiellen Einsatzmöglichkeiten entsprechend sind auch die praktischen Anwendungen des Konzeptes der seismischen Stratigraphie breit gefächert. In den zahlreichen Publikationen ist eine große Heterogenität festzustellen. Das bezieht sich nicht nur auf die jeweilige geologische Aufgabenstellung oder die zur Verfügung stehenden A-priori-Daten, sondern gleichermaßen auf das Erkundungsstadium des Einsatzes und die verwendete Analysenmethodik. Bei der Literaturdurchsicht fiel darüber hinaus besonders auf, daß eine größere Zahl von Veröffentlichungen im Titel zwar den modernen Begriff „seismische Stratigraphie" benutzt, es dann aber bei sehr allgemeinen Bemerkungen zu dieser Arbeitsrichtung bewenden läßt. Diese Arbeiten werden bei den folgenden Ausführungen nicht berücksichtigt.

In der Diskussion der praktischen Anwendungen wird kein Gesamtüberblick über die vorhandene Literatur gegeben, sondern es werden spezielle Aspekte der Anwendungsmethodik und neue Lösungswege herausgestellt sowie durch ausführlichere Erläuterung verschiedener praktischer Beispiele ergänzt.

Qualitative seismostratigraphische Gliederung

Zur klassischen Aufgabe der seismischen Stratigraphie – der Gliederung des Profils in seismische Fazieseinheiten und der Ableitung von Aussagen über Sedimentationsmilieu, geologische Entwicklungsgeschichte und Kohlenwassertoff-Höffigkeit – finden sich erste Anwendungsbeispiele bereits im schon genannten AAPG-Memoir Nr. 26. Sehr ausführlich werden dort Regionaluntersuchungen auf dem brasilianischen Schelf (BROWN & FISHER 1977) sowie im Golf von Mexiko (STUART & CAUGHEY 1977) erläutert. WEIMER & DAVIS (1977) beschreiben komplexe Untersuchungen unter Einbeziehung seismostratigraphischer Analysen zur Trassierung von Störungszonen im Denver-Becken. In der umfassenden Arbeit von VAIL u. a. (1977) werden u. a. praktische Anwendungen der seismostratigraphischen Gliederung in mesozoischen Sedimenten (in Teil 8) sowie im klastischen und karbonatischen Sedimentationsmilieu (in Teil 9 bzw. 10) diskutiert.

Weitere Beispiele unter karbonatischen Sedimentationsbedingungen stammen von sowjetischen Spezialisten, die eine Ausgliederung von Riffkörpern im Kaspischen Meer (MAMEDOV 1986) und im Schwarzen Meer (IZMAGILOV & BOVKUN 1986) versuchten. Einzelne Arbeiten befassen sich auch mit der faziellen Gliederung und der Morphologie von Evaporiten (NELY 1980) oder der seismostratigraphischen Analyse von Kohlelagerstätten in Kanada (LAWTON 1985) bzw. in Österreich (WEBER & SCHMÖLLER 1986).

Besondere Aktivität auf dem Gebiet der seismischen Stratigraphie haben ungarische Geophysiker entwickelt, die in zahlreichen Veröffentlichungen Beispiele für die seismostratigraphische Interpretation von känozoischen Sedimenten des Pannonischen Beckens liefern, und zwar vor allem im Rahmen von Regionalarbeiten mit nur geringem Bohraufschlußgrad (z. B. KESMÁRKI, POGÁCSÁS & SZANYI 1981; POGÁCSÁS 1984), teilweise auch in enger Zusammenarbeit mit Spezialisten aus den USA (MATTICK, RUMPLER & PHILLIPS 1985).

Als wichtiges Hilfsmittel der seismostratigraphischen Analyse wird die Darstellung sogenannter Paläozeitschnitte angesehen (z. B. GERASIMOVIČ 1983; GOGONENKOV u. a. 1984). Darunter ist eine Wiedergabe des seismischen Zeitschnittes zu verstehen, bei der für ein definiertes stratigraphisches Niveau (gewöhnlich für einen Leithorizont) eine statische Korrektur auf gleiche Reflexionslaufzeit vorgenommen wird. Derartige "eingeebnete" Schnitte sind besonders für die Rekonstruktion der paläomorphologischen Entwicklungsgeschichte des Untersuchungsraumes von Bedeutung.

Ein informatives Anwendungsbeispiel der seismischen Stratigraphie für die Vorhersage des Sedimentationsmodells im Zentralgebiet der Adria wird von TUFEKČIĆ (1978) gegeben. Im Meßgebiet ist ein Profilnetz vermessen worden (Abb. 10). Durch Erfassung und Verfolgung von zwei Diskordanzflächen sowie der zwischen beiden Flächen vorhandenen Reflexionsstruktur war es möglich, ein ca. 60 km langes, etwa NE–SW streichendes Paläotal nachzuweisen, das nach Südwesten abtaucht und sich verbreitert. Seine maximale Mächtigkeit beträgt etwa 2000 m. In Abb. 11 sind die Zeitschnitte dreier ausgewählter Profile dargestellt, die die Begrenzung des Paläotales deutlich erkennen lassen. Durch Betrachtungen über Reflektorkrümmung und Mächtigkeitsverhältnisse war es auf Profilen im SW-Teil des Meßgebietes möglich, Kanäle innerhalb

Eingang des Manuskripts in der Redaktion: 18. 6. 1987

[1]) Z. angew. Geol. 34 (1988) 11, S. 321–325

Abb. 10. Beispiel für die seismische Faziesanalyse, nach TUFEKČIĆ (1978)
Lage des Meßgebietes

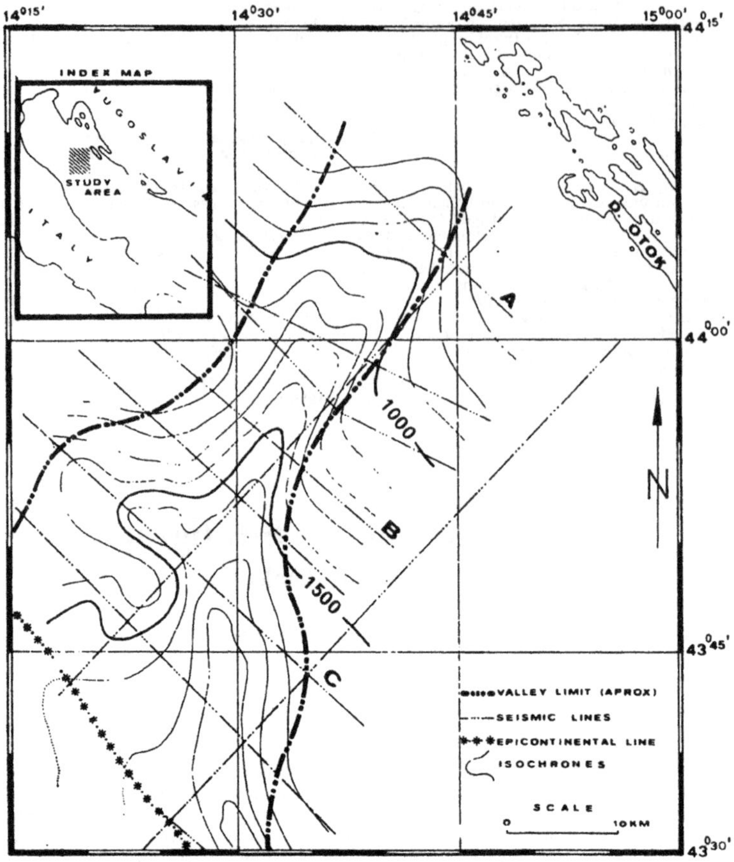

Abb. 11. Beispiel für die seismische Faziesanalyse, nach TUFEKČIĆ (1978)
Zeitschnitte für verschiedene Profile
a – Profil A
b – Profil B
c – Profil C, NW-Teil
d – Profil C, SE-Teil

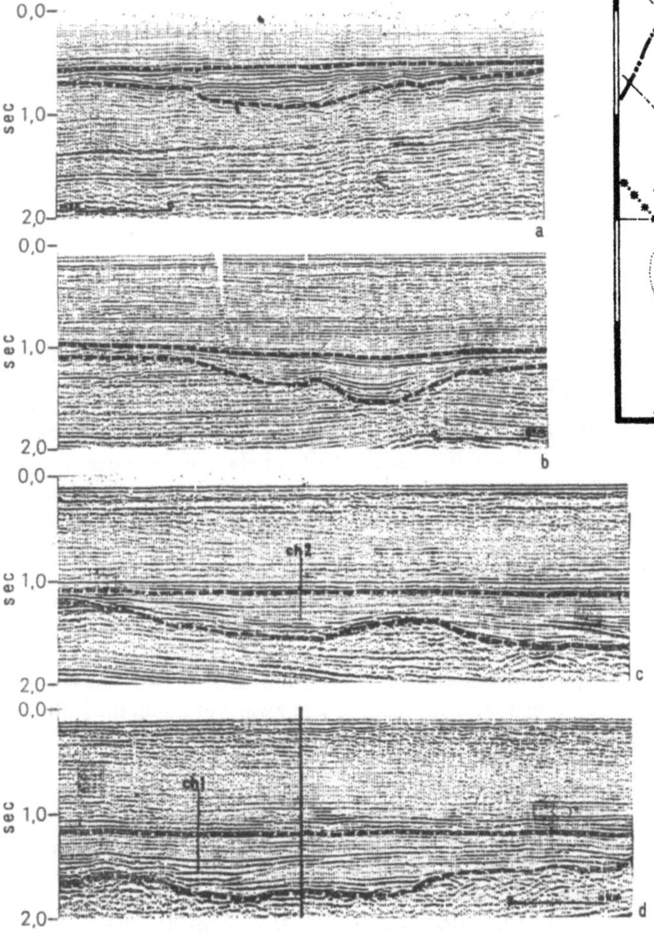

des Paläotales auszugliedern (ch1 und ch2 in Abb. 11), die wahrscheinlich Linsen grobkörniger Sandsteine darstellen. Ihre z. T. vorhandene Amplitudenüberhöhung wird als Ausdruck einer möglichen Kohlenwasserstoff-Führung angesehen.

Nutzung der seismischen Modellierung

Ein wichtiges Hilfsmittel bei der Interpretation, insbesondere bei Aufgaben der Detailerkundung, stellt die seismische Modellierung dar. Nach MECKEL & NATH (1977) lassen sich dabei drei Anwendungsschwerpunkte unterscheiden:

1. Einschätzung der Erfolgsaussichten für die Lösung einer geologischen Aufgabenstellung vor Meßbeginn,
2. Analyse der Kongruenz verschiedener Modellvarianten mit dem realen Meßergebnis,
3. Untersuchungen über die mögliche Ursache von Anomalien im seismischen Wellenfeld bei ungenauen geologischen Modellvorstellungen.

Die ersten bekannten Arbeiten befassen sich naturgemäß vor allem mit dem erstgenannten Anwendungsbereich. Als erste veröffentlichten HARMS & TACKENBERG (1972) Untersuchungen über die Widerspiegelung typischer Sedimentationsbedingungen im seismischen Wellenbild. Für drei Sedimentationsmodelle von Sand/Tonfolgen wurden synthetische Zeitschnitte berechnet und versucht, die Modelle anhand der Reflexionspolarität, der flächenhaften Variation der Reflexionsamplitude, der Persistenz der Reflexionseinsätze und der Intervallgeschwindigkeit zu unterscheiden.
Analoge Untersuchungsergebnisse legten später auch KHATTRI & GIR (1976) sowie KHATTRI, SINVHAL & AWASTHI (1979) vor. Dabei wurden auch spezielle Kenngrößen des Leistungsspektrums und der Autokorrelationsfunktion hinsichtlich ihres Aussagewertes für eine lithostratigraphische Unterscheidung getestet. Analysiert wurden vier Modelle mit zyklischer Sedimentation: Sand/Ton- bzw. Kohle/Ton-Wechsellagerung, Transgressions/Regressionsmodell von Küstenlinien und Basalsandmodell. TAL'VIRSKIJ u. a. (1985) untersuchten ebenfalls die Unterscheidungsmöglichkeiten — und zwar zwischen litoralem und neritischem Sedimentationsmilieu — anhand spezieller dynamischer Parameter des Wellenfeldes (mittlere und maximale Amplitude, mittlere Periode, Verhältnisse zwischen benachbarten Halbschwingungen usw.).
GALLOWAY, YANCEY & WHIPPLE (1977) erläutern an einem praktischen Beispiel, wie unter Verwendung der seismischen Modellierung die Entwicklungsgeschichte des Tafelrandes im E-Teil des Anadarko-Beckens rekonstruiert und drei Bildungsetappen ausgegliedert werden konnten.
Bei praktischen Untersuchungen zur Erfassung von Erosions-

rinnen im Gebiet Saratov wurden zur Interpretation der Meßergebnisse von GUTERMAN u. a. (1984) ebenfalls Analysen von synthetischen Seismogrammen herangezogen.

Unter der in der westlichen Literatur verbreiteten Bezeichung „Stratigraphische Modellierung und Interpretation" verbirgt sich eine Arbeitsrichtung, die sich vornehmlich mit der Detailuntersuchung höffiger oder produktiver Bereiche unter Zuhilfenahme der seismischen Modellierung befaßt. Wesentliche Arbeiten hierzu wurden von verschiedenen Mitarbeitern der Firma GeoQuest International aus Houston, Texas, im AAPG-Memoir Nr. 26 veröffentlicht und von NEIDELL (1979) in einer Monographie zusammengefaßt. Sehr detailliert sind hier die Nachweisbedingungen für verschiedene lithostratigraphische Modelle (Kanalsande, litorale Barrensande, Riffkörper usw.) analysiert worden, bis hin zum Einfluß eines Wechsels der Porenfüllung im Speicher. Praktische Anwendungsbeispiele für die Einbeziehung der seismischen Modellierung in die Detailerkundung von Lagerstätten stammen u. a. von RAFFALOVICH & DAW (1984) sowie von FOCHT & BAKER (1985).

Eine wichtige Rolle spielt bei solchen Untersuchungen die Ermittlung der Mächtigkeit von Speichern, deren Erfassung an der Grenze der seismischen Auflösung liegt. Ausgangspunkt der Methodik ist die Tatsache, daß die Bestimmung der Mächtigkeit einer Schicht (mit gegenüber dem umgebenden Material veränderter Schallhärte) über die Laufzeitdifferenz zwischen dem entsprechenden Maximum und Minimum nur bis zu einer Mächtigkeit $\geqq 1/2$ Periodenlänge möglich ist und hier — durch konstruktive Interferenz zwischen der Ober- und Unterkantenreflexion bedingt — eine erhöhte Reflexionsamplitude zu beobachten ist (Abb. 12). Diese Beziehungen zwischen tatsächlicher und scheinbarer Mächtigkeit sowie der Signalamplitude im Bereich der sogenannten Tuningmächtigkeit einer auskeilenden seismischen Schicht sind schematisch in Abb. 13 dargestellt. Aus der graphischen Darstellung wird deutlich, daß auch unterhalb der Tuningmächtigkeit seismischer Schichten eine quantitative Mächtigkeitsbestimmung über die Reflexionsamplitude möglich ist.

Praktisch wird so vorgegangen, daß unter Verwendung des aus den realen Daten extrahierten Signals (oder eines dem verfügbaren Frequenzbereich entsprechenden theoretischen Signals) synthetische Spuren für unterschiedliche Schichtmächtigkeiten berechnet und daraus Eichamplitudenkurven bestimmt werden.

Die Methodik wird vergleichsweise häufig benutzt für die Mächtigkeitsbestimmung geringmächtiger höffiger oder produktiver Sandsteinspeicher (z. B. BEITZEL & ROE 1981; COSSEY & VALENTA 1984) bzw. zur Korrektur von Tuningeffekten bei der quantitativen Auswertung von Amplitudenanomalien im Bereich von Gassanden (BROWN, DAHM & GREBNER 1984, 1986), teilweise auch bereits zur Kontrolle des Füllstandes von Untergrundgasspeichern (SCHMOLL & HELBIG 1982) oder des Förderfortschrittes bei Gaslagerstätten (GREAVES, FULP & HEAD 1984) durch Auswertung von Wiederholungsmessungen.

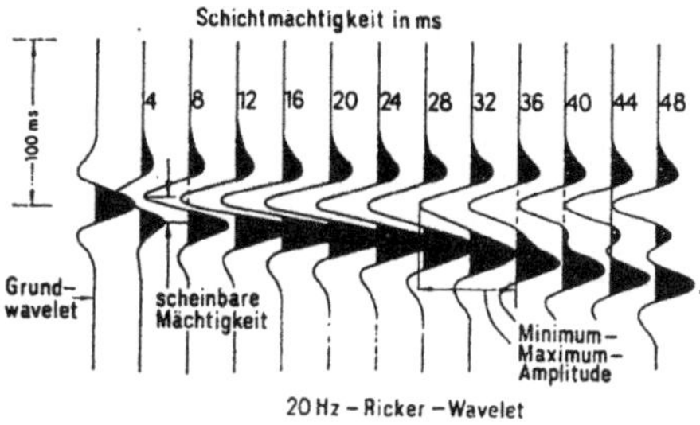

Abb. 12. Synthetische Spuren für eine auskeilende Schicht, nach NEIDELL (1979)
Inputsignal: 20-Hz-Ricker-Wavelet

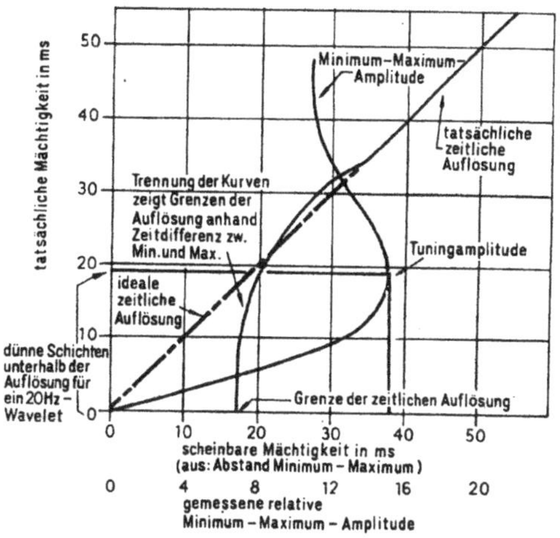

Abb. 13. Beziehungen zwischen tatsächlicher und scheinbarer Mächtigkeit einer dünnen Schicht sowie der Reflexionsamplitude für ein 20-Hz-Ricker-Wavelet, nach NEIDELL (1979)

Einbeziehung von Bohrlochmeßkurven

Außer bei der seismischen Modellierung werden Bohrlochmeßdaten auch für die Analyse der faziellen Situation im Untersuchungsgebiet herangezogen. Dabei wird die Tatsache genutzt, daß sich typische Sedimentationsbedingungen in charakteristischen Kurvenformen der Bohrlochmessung widerspiegeln. Als Ausgangsmaterial sind nach GOGONENKOV u. a. (1984) Eigenpotentialmessungen besonders geeignet, darüber hinaus auch $\gamma\gamma$- und γ-Logs. Durch spezifizierte Analyse der Kurvenform unter Einbeziehung insbesondere von Korngrößenuntersuchungen wird die sogenannte bohrlochgeophysikalische Fazies bestimmt und für die Präzisierung des lithofaziellen Ausgangsmodells benutzt. Für klastisches Sedimentationsmilieu ist die Methodik gut ausgearbeitet und wird vor allem bei Untersuchungen zur Genese von Sandkörpern eingesetzt.

Darüber hinaus werden Bohrlochmeßkurven auch direkt in die seismostratigraphische Interpretation einbezogen. Hierfür ist eine Transformation der original im Teufenmaßstab vorliegenden Kurven in den seismischen Zeitmaßstab, gegebenenfalls eine Filterung entsprechend dem seismischen Nutzfrequenzspektrum und eine Projizierung dieser Meßkurven in den Stapelzeitschnitt oder in spezielle Parameterverteilungen erforderlich.

Aushaltung von Sedimentationszyklen über Spektralanalyse

Für diese Aufgabenstellung wird von BRODOV & MUŠIN (1985) ein interessanter Lösungsweg vorgeschlagen. Ausgangspunkt sind charakteristische Abfolgen in der Sedimentablagerung, die Sedimentationszyklen, die systematische und kontinuierliche Änderungen der stofflichen (und strukturellen) Eigenschaften der Sedimentfolge aufweisen und die in der Regel durch Sedimentationsunterbrechungen begrenzt werden. Im klastischen Profil lassen sich vier Grundtypen von Sedimentationszyklen unterscheiden, die durch entsprechende Änderungen in der Korngrößenzusammensetzung unterschieden werden. Diese Grundtypen (I — Prozyklus, II — Rezyklus, III — Reprozyklus, IV — Prorezyklus) sind in Abb. 14, Spalte 1 in entsprechender Symbolik dargestellt.

Es wird davon ausgegangen, daß diese Sedimentationsmodelle eine spezifische petroakustische Charakteristik aufweisen: feinkörnige Schichtenfolgen weisen im Mittel eine geringere Schichtmächtigkeit, einen kleineren Schallhärtewert und eine

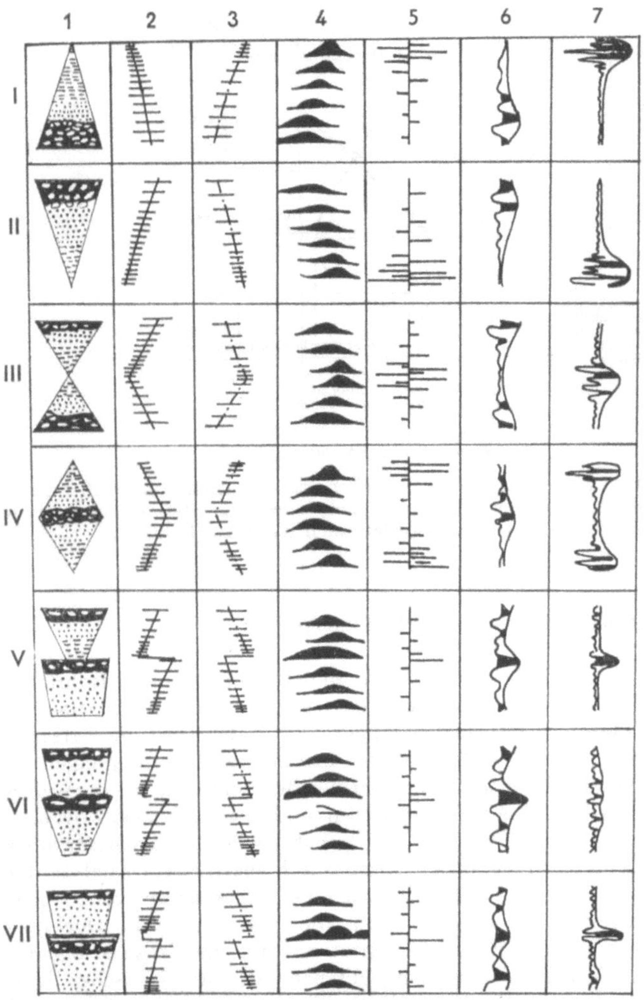

Abb. 14. Schematische Klassifizierung von Sedimentationszyklen, ihnen entsprechende Änderungen der petroakustischen Eigenschaften sowie deren Widerspiegelung im seismischen Wellenbild, nach BRODOV & MUŠIN (1985)

1 – Zyklen- und Hiatustypen I–VII, 2 – Schallhärte und ihre Dispersion, 3 – Häufigkeit des Schichtwechsels und ihre Dispersion, 4 – Leistungsspektren, 5 – Reflexionskoeffizientenfolge, 6 – synthetische seismische Spur bei niederfrequenter Filterung, 7 – synthetische seismische Spur bei hochfrequenter Filterung

höhere Schallhärtedifferenzierung an den Schichtgrenzen auf (also höhere Reflexionskoeffizienten) als grobkörnigere Sedimente. Diese spezifischen Änderungen des petroakustischen Profils innerhalb des jeweiligen Sedimentationszyklus erzeugen charakteristische Signaturen im seismischen Wellenbild (Abb. 14, Spalten 6 u. 7).

Die Aushaltung der entsprechenden seismischen Zyklen erfolgt über schmalbandige Filterungen von Zeitabschnitten mit unterschiedlichem Durchlaßbereich bzw. gleitende Spektralanalyse mit unterschiedlicher Länge des Analysefensters. Je nach Größe der Glättungsbasis werden seismische Zyklen verschiedener hierarchischer Ordnung (Mikro-, Meso- oder Makrozyklen) ausgehalten.

KLUŠIN & TAL'VIRSKIJ (1986) befassen sich mit den Grundlagen der Methodik anhand theoretischer Überlegungen über die spektrale Widerspiegelung verschiedener Schichtmodelle und schlagen eine (spiegelsymmetrische) Darstellung der Ergebnisse der gleitenden Spektralanalyse nicht in Abhängigkeit von der Frequenz f sondern von deren Reziprokwert $\Delta t = 1/2f$ vor. Das Konzept wird anhand entsprechender Darstellungen aus zwei verschiedenen Meßgebieten und deren Interpretation ausführlich erläutert.

Das erläuterte Konzept ist nur im rein klastischen Schichtprofil gültig. Abweichungen hiervon, wie beispielsweise evaporitische Einlagerungen, stellen Störeffekte dar. Grundvoraussetzung für die Anwendung ist in jedem Fall seismisches Ausgangsmaterial sehr guter Qualität und mit breitem Nutzfrequenzspektrum. Inwieweit das erläuterte Konzept tatsächlich in größerem Umfang praktisch genutzt werden kann, hängt darüber hinaus wesentlich von der Gültigkeit der postulierten Beziehungen zwischen Zyklus, petroakustischem Profil und seiner Widerspiegelung im seismischen Wellenbild ab und kann derzeit nur schwer beurteilt werden.

Automatische Ausgliederung und quantitative Analyse von Fazieseinheiten

Für die anspruchsvolle Aufgabenstellung der maschinellen Untergliederung des Zeitschnittes in seismische Fazieseinheiten existiert bisher nur wenig Literatur. Nach ersten Versuchen von HARJES (1979) in dieser Richtung stellte CONTICINI (1985) ein interessantes Lösungskonzept vor. Ausgangsdaten sind spezielle aus dem seismischen Zeitschnitt extrahierte Parameter des Wellenfeldes (Intervallgeschwindigkeit, Amplitude, Frequenz, Persistenz). Mit Hilfe statistischer Verfahren (u. a. Clusteranalyse) werden zunächst seismische Referenzfazies bestimmt, die durch die Mittelwerte und Streuungen der Wellenfeldparameter charakterisiert sind. Die räumliche Ausdehnung jeder Fazieseinheit wird durch mehrdimensionale Ähnlichkeitsanalyse ermittelt. Die Darstellung der Endergebnisse erfolgt als Farbschnitte, in denen unterschiedliche seismische Fazies markiert sind und die Farbintensität ein Maß für die Ähnlichkeit mit der jeweiligen Referenzfazies darstellt.

Versuche zum Einsatz statistischer Bearbeitungsverfahren für die Erfassung und Quantifizierung lithofazieller Änderungen oder auch eines Wechsels der Porenfüllung innerhalb definierter Teufenabschnitte sind demgegenüber nicht neu. Das klassische Beispiel hierfür stellen die Arbeiten von MATHIEU & RICE (1969) zur Ermittlung der Vertonungsgrenze eines sandigen Speichers mit Hilfe der Diskriminanzanalyse dar. In der neueren Literatur wird für solche Aufgabenstellungen neben diesen Verfahren (SINVHAL, KHATTRI u. a. 1984) auch die Methode der Modellerkennung (BOIS 1980, 1982) sowie die Clusteranalyse (HUANG & FU 1984) herangezogen.

In diese Arbeitsrichtung gehören auch die ersten Untersuchungen auf dem Gebiet der DDR zur Nutzung der Modellerkennung für die Erfassung eines markanten Fazieswechsels im sedimentären Rotliegenden (PRÖHL, PATZER u. a. 1986).

Nutzung von 3D-Messungen

Durch Verwendung von 3D-Daten als Ausgangsmaterial eröffnen sich neue Möglichkeiten für die seismostratigraphische Interpretation.

Bei Analysen im *Profilschnitt* können Besonderheiten des Wellenfeldes entlang von Profilen mit praktisch beliebiger Richtung sowie ihre Änderungen von Profil zu Profil sehr detailliert verfolgt werden. Ausgangsmaterial für die Interpretation sind dabei Darstellungen der seismischen Zeitschnitte selbst (häufig mit verschiedenen Polaritäten oder durch farblich unterschiedlich gekennzeichnete positive und negative Halbschwingungen) oder die Verteilung in der x-t-Ebene von speziellen aus den Daten bestimmten Kenngrößen, wie momentane dynamische Parameter (Amplitude, Frequenz, Phase) oder die über Spurinversion gewonnenen Pseudogeschwindigkeiten.

Insbesondere die flächenhafte Darstellung der Verteilung bestimmter Kenngrößen (derzeit vor allem der Reflexionsamplitude) in sogenannten *Zeitscheiben*, und zwar entsprechend einer konstanten Reflexionslaufzeit oder im Laufzeitniveau eines interessierenden Horizontes, ermöglicht eine völlig neue Qualität der Interpretation.

Als klassisches Beispiel für den erreichbaren Informationsgewinn kann die Veröffentlichung von Brown, Dahm & Graebner (1981) angesehen werden. Die entsprechenden Messungen erfolgten zur Erkundung eines durch drei Bohrungen angetroffenen Gasfeldes im Golf von Thailand. Dabei wurde eine Fläche von 11 × 13 km² mit einem Profilabstand von 100 m und einem Tiefenpunktabstand von 33 m übermessen. Geologisch handelt es sich hier um 3 km mächtige tertiäre Sedimente, die in drei Hauptzyklen unter größtenteils randlichen marinen Bedingungen mit kontinentaler Beeinflussung abgelagert wurden. Die spezifizierte Analyse des Datenmaterials, insbesondere in der Darstellung als Zeitscheiben, zeigt, daß für das Meßgebiet Deltabildungen mit Barren- und Rinnensanden charakteristisch sind, die bei stark wechselnden Mächtigkeiten eine komplizierte und vielgliedrige räumliche Verteilungsstruktur aufweisen.

Auf den Abb. 15 und 16 sind als Beispiel zwei Zeitscheiben dargestellt, die die Reflexionsamplituden-Verteilung für eine jeweils konstante Reflexionslaufzeit veranschaulichen (im Original farbig kodiert). Abb. 15 zeigt, mit welcher Detailliertheit der nach Norden gerichtete Schüttungskegel eines Deltas und eine dem Delta vorgelagerte Barre, die sich in nordwestlicher Richtung auffächert, erfaßt werden können. Auf Abb. 16 ist schließlich ein stark mäandrierender fossiler Flußlauf im Bereich der Deltaebene in allen Einzelheiten verfolgbar. Der Deltahang selbst deutet sich hier durch eine Aufgabelung im NW-Teil der Meßfläche an.

Weitere 3D-Anwendungsbeispiele stammen aus dem Golf von Mexiko, wo vor der Küste von Louisiana für die Detailerkundung produktiver pleistozäner Gassande Messungen mit 50 m Profilabstand und 12,5 m Tiefenpunktabstand vorgenommen wurden (Brown, Dahm & Graebner 1984, 1986). Durch eine speziell angepaßte Auswertetechnik, die Bright-Spot-Effekte berücksichtigte und eine entsprechende Eichung an bereits vorhandenen Bohrungen einschloß, konnten sehr detaillierte Karten der produktiven Mächtigkeit von zwei lateral in ihrer Mächtigkeit und Gasführung sehr variablen Speicherhorizonten vorgelegt werden.

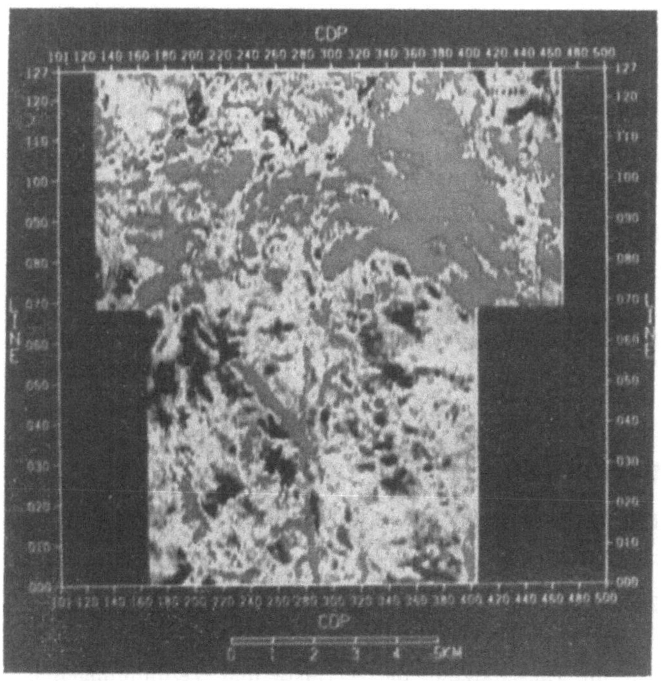

Abb. 15. Darstellung der Reflexionsintensität für eine Zeitscheibe bei 304 ms, aus Brown, Dahm & Graebner (1981)

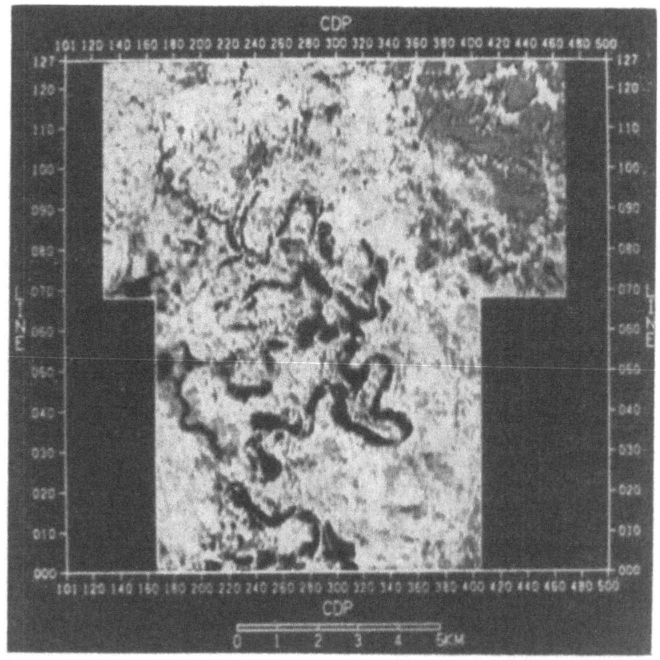

Abb. 16. Darstellung der Reflexionsintensität für eine Zeitscheibe bei 196 ms, aus Brown, Dahm & Graebner (1981)

Voraussetzungen für die praktische Anwendung

Aus den beschriebenen Beispielen wird deutlich, daß es für die Untersuchungen zur seismischen Stratigraphie keine allgemeingültige Bearbeitungs- und Interpretationsmethodik gibt und auch die Anforderungen an das Ausgangsmaterial und die Zusatzinformationen — in Abhängigkeit von der geologischen Aufgabenstellung und dem Einsatzstadium — sehr unterschiedlich ausfallen. Aus diesem Grunde lassen sich keine allgemeinen Vorschriften für die praktische Anwendung des Konzeptes der seismischen Stratigraphie formulieren. Die im folgenden genannten Forderungen müssen deshalb auch nicht in jedem Falle voll erfüllt sein, sie ergeben in ihrer Summe die optimalen Voraussetzungen für praktische seismostratigraphische Untersuchungen.

Geologisches Modell

Es sind sämtliche notwendigen, relevanten, aus Bohrungen verfügbaren Ausgangsdaten mit dem Ziel der Erarbeitung eines geologischen Modells zusammenzustellen. Dazu zählen Angaben zur Stratigraphie, Lithologie, Speichercharakteristik, Tektonik bis hin zur Beschreibung von Kern- und Spülproben ebenso, wie alle verwendbaren bohrlochgeophysikalischen Meßergebnisse (s. u.). Das geologische Modell bildet eine wesentliche Voraussetzung für die Interpretation. Zu erarbeiten ist ein komplexes Modell, in dem nicht nur die strukturellen, sondern auch die lithofaziellen Verhältnisse des Untersuchungsraumes zu berücksichtigen sind.

Qualität des seismischen Ausgangsmaterials

Die Anforderungen an die Qualität des seismischen Stapelzeitschnittes sind wesentlich höher als bei reiner struktureller Interpretation. Praktische Anwendungen erfolgten fast ausnahmslos an Material mit sehr guter Datenqualität. Gogonenkov u. a. (1984) formulieren als Mindestforderung für die seismostratigraphische Detailerkundung die Existenz einer reellen Reflexionsphasenachse auf allen Spuren des Einzelseismogramms und ein Nutz/Stör-Verhältnis der Endstapelung von 5...20.

Meßdichte

Die Dichte des Meßnetzes muß größer sein als bei der konventionellen Strukturvorbereitung. Für die Analyse der äußeren Form seismischer Fazies- (oder Subfazies-) Einheiten ist eine sichere flächenhafte Trassierung wesentlicher Besonderheiten des Wellenfeldes erforderlich. Am günstigsten hierfür sind echte 3D-Messungen mit regelmäßigem Meßgitter.

Geophysikalische Bohrlochmessungen

Durch gezielte Einbeziehung von Bohrergebnissen kann die Qualität der seismostratigraphischen Interpretation wesentlich verbessert werden. Dabei handelt es sich vor allem um die folgenden beiden Aufgaben:

- Erarbeitung des seismoakustischen Modells (im wesentlichen aus Akustiklog, Dichtelog und Ergebnissen der seismischen Vertikalprofilierung) für die Einbeziehung seismischer Modellierungsarbeiten in die Interpretation.
- Analyse der bohrlochgeophysikalischen Fazies (aus SP-, γ-, $\gamma\gamma$-Log, granulometrischen Daten usw.) für die Stützung der seismischen Faziesanalyse.

Seismische Grundbearbeitung

Wegen der hohen Anforderungen an die Datenqualität und der häufig geforderten quantitativen Aussagen ist im allgemeinen eine Datenbearbeitung mit wahren Amplituden, speziell angepaßter (Nullphasen-) Dekonvolution sowie eine Stapelung mit hochgenauen statischen und kinematischen Korrekturen notwendig.

Seismische Spezialbearbeitung

Neben dem Zeitschnitt selbst werden zunehmend verschiedene aus den seismischen Daten extrahierte Parameterverteilungen in die Faziesanalyse einbezogen. Hierzu zählen insbesondere die über Hilbert-Transformation berechneten momentanen dynamischen Parameter und die aus der Spurinversion resultierenden Pseudogeschwindigkeiten, daneben auch verstärkt Ergebnisse der Spektralanalyse und der maschinellen Komplexinterpretation verschiedener Parameterverteilungen.

Ergebnisdarstellung

Der geeigneten Darstellung der zu interpretierenden Daten kommt mehr und mehr eine Schlüsselstellung zu. Voraussetzung für eine adäquate Ausschöpfung des Informationsgehaltes, insbesondere bei Einbeziehung verschiedener aus dem Zeitschnitt extrahierter Parameterverteilungen und von Bohrdaten, ist die Nutzung moderner Bildverarbeitungssysteme mit ihren vielfältigen Möglichkeiten zur farbigen Datenwiedergabe.

Schlußfolgerungen

Die Auswertung der internationalen Literatur zeigt, daß weltweit intensiv an der Weiterentwicklung und Anwendung der seismischen Stratigraphie gearbeitet wird. Dabei wird deutlich, daß hinsichtlich der Anforderungen an das Ausgangsmaterial und der eingesetzten Bearbeitungsverfahren weitgehend Analogien zum Konzept der seismischen Substanzaussage bestehen. Der qualitative Unterschied besteht in der Einbeziehung zusätzlicher spezifischer Interpretationskriterien, die zur Ausgliederung sogenannter seismischer Faziesseinheiten führt.

Das bei praktischen Untersuchungen eingesetzte Verfahrensinventar und die benutzte Interpretationsmethodik sind außerordentlich unterschiedlich: Sie reichen von einfacher visueller Analyse des Zeitschnittes bis zum komplexen Einsatz einer breiten Palette von Spezialverfahren zur Datenbearbeitung, -aufbereitung und -darstellung unter Einbeziehung verschiedenster Zusatzinformationen, von der Analyse von Regionalprofilen bis zur Bearbeitung eines engmaschigen Meßnetzes in Lagerstättenbereichen.
Dementsprechend breit gefächert sind auch die prinzipiellen Aussagemöglichkeiten: Sie reichen von allgemeinen Hinweisen zur Rayonierung oder zur Höffigkeitseinschätzung über Aussagen zum Sedimentationsmilieu, zu Art und Richtung des Sedimenttransportes und zur geologischen Entwicklungsgeschichte bis zu Detailangaben über die Lagerstättenkontur, zu den Speichereigenschaften oder zur Produktivität von bereits durch Bohrungen aufgeschlossenen Erdöl- und Erdgas-Lagerstätten.
Für die Gewinnung solcher spezifizierter lithofazieller Aussagen sind eine enge interdisziplinäre Zusammenarbeit insbesondere von Seismikern und Geologen, und zwar in allen Bearbeitungsetappen von der Projektierung bis zur Endinterpretation ebenso zwingend erforderlich wie große Erfahrungen im Umgang mit dem Interpretationskonzept und detaillierte Vertrautheit mit den Besonderheiten des Meßgebietes. Die Anforderungen an das Ausgangsmaterial sind — ebenso wie bei Arbeiten zum seismischen Direktnachweis — wesentlich größer als für die konventionelle Strukturerkundung. Voraussetzungen sind eine höhere Meßdichte bei Detailarbeiten, erheblich bessere Datenqualität, höherer Aufschlußgrad durch Bohrungen und die geeignete Aufbereitung und Bereitstellung aller verfügbaren Zusatzinformationen.

Zusammenfassung

Im Ergebnis einer Auswertung der internationalen Literatur wird die neue Arbeitsrichtung seismische Stratigraphie vorgestellt. Dabei werden zunächst die wesentlichen Grundbegriffe und das zugrundeliegende methodische Konzept erläutert. Die vielfältigen Möglichkeiten der praktischen Nutzung werden anhand von Anwendungsbeispielen diskutiert.

Резюме

На основании литературного обзора представлено новое научно-прикладное направление — сейсмическая стратиграфия. В начале работы разъясняются основные понятия этого направления и лежащая в его основе методическая программа. На конкретных примерах рассматриваются многочисленные возможности практического использования метода.

Summary

Resulting of an analysis of international literature the new subject of seismic stratigraphy is presented. The most important basic principles and the methodical conception is explained. The manyfolded possibilities of its practical use are discussed by means of some examples.

Die Ableitung optimaler Erkundungsnetze auf der Grundlage einer geostatistischen Einschätzung der Verluste und Verdünnungen

Joachim Menz, Freiberg

In dem folgenden Beitrag werden Ergebnisse mitgeteilt, die aus der Zusammenarbeit des Wissenschaftsbereiches Markscheidewesen der Bergakademie Freiberg mit dem Lehrstuhl für Markscheidewesen des Moskauer Bergbauinstitutes resultieren.

Eingang des Manuskripts in der Redaktion: 30. 3. 1987

Zulässigkeit des Analogieschlusses — Voraussetzung für die Anwendung des Verfahrens

Für die Einschätzung der Verluste und Verdünnungen benötigt man ein repräsentatives Gebiet eines bereits abgebauten und damit ausreichend bekannten Lagerstättenteils. Repräsentativ bedeutet einerseits, daß die markscheiderisch-geologischen Daten der Erkundung und Abbaubemusterung repräsentativ die Gesetzmäßigkeiten der Lagerstättenparameter in dem Gebiet widerspiegeln. Repräsentativ heißt andererseits, daß die im Gebiet festgestellten Gesetzmäßigkeiten für die gesamte Lagerstätte repräsentativ sind und auf die noch zu erkundenden Lagerstättenteile übertragen werden können. Die Zulässigkeit dieses Analogieschlusses ist Voraussetzung für die Anwendung des Verfahrens. In einem geologischen Gutachten muß unter Einbeziehung aller Vorkenntnisse und unter Berücksichtigung aller Erkundungsdaten dazu eine Einschätzung gegeben werden.

Prüfung der Zulässigkeit des Analogieschlusses mit der Varianzanalyse

Liegen bereits Daten der Detailerkundung in den noch weiter zu erkundenden Gebieten vor, kann die Varianzanalyse zur Entscheidungsfindung mit herangezogen werden. Bei Anwendung der Varianzanalyse werden die Erkundungsdaten eines Lagerstättenparameters in Gruppen unterteilt. Die Gruppierung richtet sich nach der Lage und Höhe der Lagerstättenbereiche. Unter der Annahme dieser beiden Einflußgrößen läßt sich das Verfahren der doppelten Varianzanalyse mit unterschiedlicher Besetzung der Gruppen nutzen, das an der Bergakademie Freiberg in einem BASIC-Programm für den Bürocomputer A 5120 aufbereitet wurde.
Bei der Varianzanalyse wird jede Gruppe als eine Stichprobe für sich angesehen. Es wird geprüft, ob die Gruppen von den Einflußgrößen unabhängig sind und damit der gleichen Grundgesamtheit zugeordnet werden können. In diesem Fall bestehen von dieser Seite her keine Einwände gegen den Analogieschluß. In sämtlichen Gruppen, so auch in dem repräsentativen Gebiet, werden nur die Daten der Detailerkundung zu dieser Untersuchung herangezogen. Die Varianzanalyse erfordert unabhängige und annähernd normalverteilte Meßwerte.

Unterscheidung verschiedener Homogenitätsbereiche nach dem Maß der Kompliziertheit

Am Lehrstuhl Markscheidewesen des Moskauer Bergbauinstitutes wird das Maß der bergbaugeologischen Kompliziertheit V_i zur Einstufung der Lagerstättenabschnitte i benutzt (BUKRINSKIJ 1985). In diesem Maß wird das gewichtete Zusammenwirken der Veränderlichkeit der für den Abbau entscheidenden Lagerstättenparameter ausgedrückt. Es wird definiert durch

$$V_i = \sum_{j=1}^{k} v_{ij} \qquad (1)$$

mit

$$v_{ij} = g_i \cdot H_j \cdot u_{ij}. \qquad (2)$$

Dabei ist g_j das Gewicht, das dem Lagerstättenparameter j auf Grund einer Expertenschätzung zugeordnet wird, die für die Auswahl der wesentlichen k Lagerstättenparameter durchzuführen ist. Die Bedeutung der einzelnen Lagerstättenparameter für den Abbau wird durch Punkte bewertet. Das Gewicht ergibt sich aus dem Verhältnis der Punkte des Parameters j zur Gesamtpunktzahl der ausgewählten k Parameter.
Die Größe H_j ist die Informationsentropie des Lagerstättenparameters j, die aus seiner empirischen Verteilung im repräsentativen Gebiet nach

$$H_j = -\sum_{l=1}^{m} p_l \cdot \log_2 p_l \qquad (3)$$

abgeleitet wird. Für p_l ist die relative Häufigkeit in der Klasse l einzusetzen. Sie ergibt sich aus

$$p_l = \frac{h_l}{n} \qquad (4)$$

wobei h_l die absolute Häufigkeit in dieser Klasse ist. Die Anzahl der Klassen der empirischen Verteilung wird nach

$$m = 5 \lg n \qquad (5)$$

berechnet, wobei n die Anzahl der Beobachtungen ist, die sich für das dichteste Erkundungsnetz im repräsentativen Gebiet ergibt. Dieses Netz soll hier als Basisnetz bezeichnet werden.
Die Größe u_{ij} drückt die Veränderlichkeit des Lagerstättenparameters j in den elementaren Lagerstättenabschnitten i aus. Werden die Netzmaschen der Detailerkundung als elementare Lagerstättenabschnitte i verwendet, dann sind die Meßwerte an den Netzpunkten in der Regel noch voneinander unabhängig. Aus den 4 Meßwerten z_{li} jeder Netzmasche i mit dem Mittelwert $\bar{z}_i$ kann in diesem Fall u_{ij} nach

$$u_{ij} = \frac{2 \sum_{li=1}^{4} |z_{li} - \bar{z}_i|}{(4-1) \cdot R} \qquad (6)$$

berechnet werden. Mit R wird die Spannweite bezeichnet, die aus allen Meßwerten im repräsentativen Gebiet ermittelt wird. BUKRINSKIJ (1985) bestimmt u_{ij} aus den Differenzen der Meßwerte. Er gibt außerdem noch Korrekturen zur Berechnung von V_i an, auf die hier nicht eingegangen werden soll. Aus der Maßeinheit bit für V_i sollen hier ebenfalls keine Schlußfolgerungen abgeleitet werden.
Wegen der Veränderlichkeit der Größen u_{ij} von Abschnitt zu Abschnitt verändert sich auch die Größe V_i entsprechend. Das Maß der bergbaugeologischen Kompliziertheit ist eine ortsabhängige Größe. Die berechneten Werte V_i werden den Mittelpunkten der entsprechenden Netzmaschen zugeordnet. Aus den geglätteten Werten $\tilde{V}_i$ werden Isolinienkarten der bergbaugeologischen Kompliziertheit sowohl für das repräsentative Gebiet als auch für die noch weiter zu erkundenden Gebiete abgeleitet. Diese Karten bilden die Grundlage für die Abgrenzung verschiedener Homogenitätsbereiche. Bereiche mit einer höheren Kompliziertheit erfordern dichtere Erkundungsnetze.

Die Ableitung optimaler Erkundungsnetze

In den Arbeiten des Lehrstuhles Markscheidewesen des Moskauer Bergbauinstitutes werden im nächsten Schritt Richtung und Betrag der minimalen und maximalen Veränderlichkeit der bergbaugeologischen Kompliziertheit bestimmt. Danach richtet sich die Orientierung der Netze und das Verhältnis ihrer Maschenweiten (BUKRINSKIJ 1985). Als Optimalitätskriterium für die Auswahl der effektivsten Netzvarianten dienen die minimalen Kosten K, die sich aus den Kosten für die betriebliche Erkundung E und dem ökonomischen Schaden S zusammensetzen, dem sogenannten „Bergbaurisiko", das als Folge der ungenauen Erkundung eintritt:

$$K = E + S. \qquad (7)$$

Zur Ermittlung dieser Kosten werden im repräsentativen Gebiet für die verschiedenen Homogenitätsbereiche verschiedene Netzvarianten untersucht.
Dabei wird davon ausgegangen, daß die Isolinienkarten der Lagerstättenparameter, die aus dem Basisnetz abgeleitet worden sind, der objektiven Realität entsprechen. Aus ihnen werden die Erkundungsdaten für die verschiedenen Netze mit systematisch vergrößerter Maschenweite im Rahmen der Simulation abgeleitet. Aus diesen „Erkundungsdaten" lassen sich wiederum Isolinienkarten für die verschiedenen Lagerstättenparameter herstellen und mit den Isolinien des Basisnetzes vergleichen. Beim Vergleich werden für alle Netzvarianten die Fehler der Vorratsbestimmung und die zu erwartenden Verluste und Verdünnungen eingeschätzt. Auf dieser Grundlage erfolgt die Kostenermittlung und die Auswahl der effektivsten Netzvarianten in den einzelnen Homogenitätsbereichen.
Die Anwendung des Verfahrens führte in der UdSSR in allen Einsatzfällen zu erheblichen Kosteneinsparungen bei der betrieblichen Erkundung. Die positiven Ergebnisse, die am Lehrstuhl Markscheidewesen des Moskauer Bergbauinstitutes erzielt werden konnten, sind Anlaß dafür, den Einsatz dieses Verfahrens unter geostatistischen Modellannahmen zu diskutieren.

Geostatistische Modellannahmen und die Modellanpassung im repräsentativen Gebiet

Die am Wissenschaftsbereich Markscheidewesen der Bergakademie Freiberg für die praktische Anwendung aufbereiteten geostatistischen Verfahren basieren auf der Modellannahme, daß die Lagerstättenparameter Z als stationäre Zufallsfunktionen $Z_s(x, \omega)$ angesehen werden können, denen in der Regel Trends $m(x)$ und Störgrößen $\xi(\omega)$ überlagert sind:

$$Z = Z_s(x, \omega) + m(x) + \xi(\omega). \quad (8)$$

Mit dem Ortsvektor x wird die Abhängigkeit vom Ort und mit den Elementarereignissen ω die Abhängigkeit vom Zufall ausgedrückt. Unter diesen Modellannahmen wurden am Wissenschaftsbereich Markscheidewesen der Bergakademie Freiberg Programme für die Ableitung des optimalen Verlaufes der Lagerstättenparameter aus den Erkundungsdaten und für die Angabe der Vorhersagefehler aufgestellt (Menz 1981, 1985, 1986).
In den verschiedenen Homogenitätsbereichen des repräsentativen Gebietes erfolgt nach dem Basisnetz zunächst die Modellanpassung durch eine Trennung der zufälligen und gesetzmäßigen Anteile in den Erkundungsdaten über einen Genauigkeitstest. Dieses Verfahren der Modellanpassung bildet die Grundlage des Programmsystems GEOSTATISTIK (Menz 1985). Durch die Modellanpassung werden die inneren Kovarianzfunktionen der Lagerstättenparameter für die unterschiedlichen Homogenitätsbereiche bestimmt, die die Grundlage für die Vorhersage und für die Ableitung des Vorhersagefehlers bilden.
Im Rahmen der Modellanpassung erhält man für alle Homogenitätsbereiche Diagramme, in denen die Vorhersagefehler für die Mitte zwischen den Erkundungspunkten in Abhängigkeit vom Abstand der Erkundungspunkte angegeben werden. Abb. 1 zeigt ein solches Diagramm.
Nach der Modellanpassung kann der optimale Verlauf der Isolinien der einzelnen Lagerstättenparameter mit dem Programmsystem GEOSTATISTIK für alle Netzvarianten der verschiedenen Homogenitätsbereiche des repräsentativen Gebietes abgeleitet und automatisch dargestellt werden. Auf der Grundlage dieser noch genaueren Isoliniendarstellungen können nun in der gleichen Weise, wie bereits oben beschrieben, die optimalen Netze über die minimalen Kosten K gefunden werden. Unter der Zulässigkeit des Analogieschlusses erfolgt die Übertragung auf die entsprechenden Homogenitätsbereiche des noch abzubauenden Gebietes.

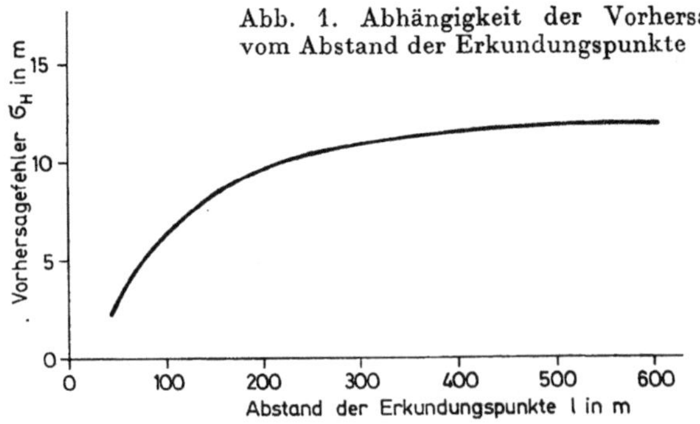

Abb. 1. Abhängigkeit der Vorhersagefehler vom Abstand der Erkundungspunkte

Neben dieser analogen Vorgehensweise, die relativ aufwendig ist, weil noch keine EDV-Programme dafür vorliegen, besteht bei der Nutzung geostatistischer Verfahren die Möglichkeit, optimale Erkundungsnetze über die Vorhersagegenauigkeit mit wesentlich geringerem Arbeitsaufwand abzuleiten. Auf dieses Verfahren, dessen Konzept bereits (Menz 1983) vorgestellt wurde, soll im folgenden näher eingegangen werden. Die Grundlage für dieses Verfahren bilden die im Rahmen der Modellanpassung für die verschiedenen Homogenitätsbereiche abgeleiteten Diagramme, in denen die Vorhersagefehler für die Mitte zwischen den Stützpunkten als Funktionen des Stützpunktabstandes dargestellt werden (Abb. 1).

Empirische Ermittlung der Verluste und Verdünnungen

In Abb. 2a wird am Beispiel des Profils einer flözförmigen Lagerstätte gezeigt, welche Verluste und Verdünnungen im Liegenden und im Hangenden auftreten, wenn im Erkundungsprofil nur noch jeder 4. Stützpunkt des Basisnetzes verwendet wird. Dabei wird zunächst angenommen, daß die technologisch-ökonomische Flözbegrenzung mit dem jeweiligen Erkundungsprofil zusammenfällt (Fall a) und daß der aus dem Basisnetz abgeleitete Flözverlauf dem in der Natur vorhandenen entspricht. Über derartige Profile aus dem repräsentativen Gebiet lassen sich die Verluste und Verdünnungen für alle zu untersuchenden Netze eines Homogenitätsbereiches ableiten und auf eine Bezugsfläche F reduzieren. Es wird erwartet, daß die empirisch ermittelten Verluste P_{ai} und Verdünnungen R_{ai} im Fall a gleich sind. Ihre Größe V_{ai} ergibt sich (bei Erwartungstreue) aus

$$V_{ai} = \frac{1}{2}(P_{ai} + R_{ai}) \quad (9)$$

für die Netzvariante i. Diese Ermittlung ist sowohl im Hangenden als auch im Liegenden notwendig.
Bei der Festlegung der Lagerstättenbegrenzung bestimmen das Abbauverfahren, die Abbauvariante und die eingesetzten Gewinnungsgeräte neben der Bergbausicherheit die technologischen Grenzen. Das dabei zu wählende Verhältnis zwischen Verlusten und Verdünnungen hängt vom Wert des Rohstoffes und von den Gehalten ab. Es ist aus ökonomischer Sicht so festzulegen, daß der größte Gewinn erzielt wird.
In Abb. 2b wird an einem vereinfachten Modell für die Hangendbegrenzung gezeigt, wie sich die Verluste und Verdünnungen verändern, wenn die technologische Lagerstätten-

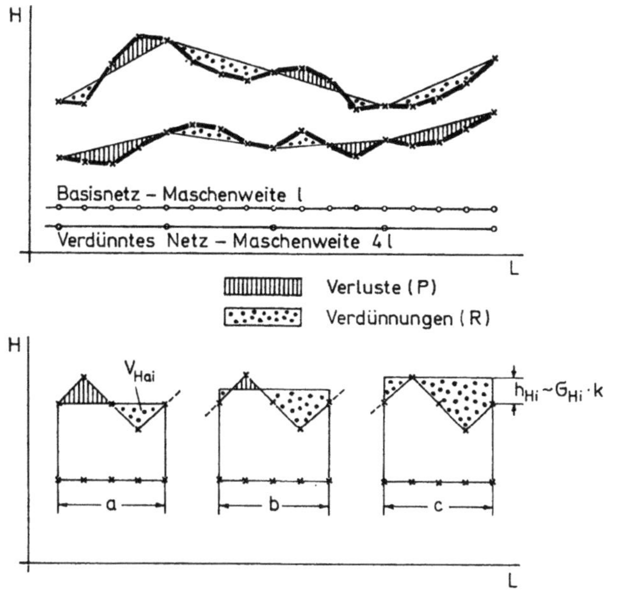

Abb. 2. Verluste und Verdünnungen durch Vergrößerung des Stützpunktabstandes im einem Erkundungsprofil

begrenzung verschoben wird. Im Intervall a wird das Modell für den Fall a vorgestellt, bei dem die technologische Grenze dem erkundeten Verlauf entspricht.

Werden die auf die Bezugsfläche F reduzierten Verluste und Verdünnungen im Hangenden für die Netzvariante i mit V_{Hai} bezeichnet, dann gilt nach dem vereinfachten Modell im Abschnitt a

$$P_{Hai} = R_{Hai} = V_{Hai} = \frac{1}{3} \cdot \frac{1}{2} F \cdot h_{Hi} \tag{10}$$

Die im Modell vorhandene maximale Abweichung h_{Hi} der Lagerstätte vom angenommenen Verlauf läßt sich damit berechnen.

Im Abschnitt b der Abb. 2b wird die technologische Grenze im Hangenden um $\frac{1}{2} h_{Hi}$ und im Abschnitt c um h_{Hi} nach oben verschoben. Für den Abschnitt b ergeben sich damit die Verluste

$$P_{Hbi} = \frac{1}{3} \cdot \frac{1}{2} \cdot \frac{1}{4} F \cdot \frac{1}{2} h_{Hi} = \frac{1}{8} V_{Hai} = 0{,}125 V_{Hai} \tag{11}$$

und die Verdünnungen

$$R_{Hbi} = \frac{1}{2} F \cdot h_{Hi} + \frac{1}{8} V_{Hai} = \frac{25}{8} V_{Hai} = 3{,}125 V_{Hai} \tag{12}$$

Im Abschnitt c erhält man schließlich die Verdünnungen

$$R_{Hci} = F \cdot h_{Hi} = 6 V_{Hai} \tag{13}$$

Verluste treten im Fall c nicht auf:

$$P_{Hci} = 0 \tag{14}$$

Erfolgt die Verschiebung nach unten, sind in den Formeln die Verluste und Verdünnungen zu vertauschen. In der gleichen Weise kann für eine Verschiebung der technologischen Abbaugrenze im Liegenden berechnet werden, wie sich die Verluste und Verdünnungen verändern. Für die praktische Anwendung kann das Modell dem Verlauf der Vorhersagefehler zwischen den Stützpunkten besser angepaßt werden.

Geostatistische Vorhersage der Verluste und Verdünnungen

Die empirische Ermittlung der Verluste und Verdünnungen nach Abb. 2a ist relativ aufwendig. Deshalb wird am Wissenschaftsbereich Markscheidewesen der Bergakademie Freiberg bei der Ermittlung der Verluste und Verdünnungen von dem Modell in Abb. 2b ausgegangen. Bei der geostatistischen Vorhersage kann die Höhe h_{Hi} im Modell aus den Vorhersagefehlern σ_{Hi} abgeleitet werden, die den in Abb. 1 gezeigten Diagrammen zu entnehmen sind. Mit σ_{Hi} werden die in der Mitte zwischen den Stützpunkten zu erwartenden Abweichungen der aus dem Netz i abgeleiteten Hangendbegrenzung von der Natur bezeichnet. Die Höhe ergibt sich aus

$$h_{Hi} \approx k \cdot \sigma_{Hi} \tag{15}$$

Der Faktor k dient zur Anpassung des tatsächlichen Verlaufes der Vorhersagefehler zwischen den Stützpunkten an das vereinfachte Modell in Abb. 2b. Ganz analog ergibt sich die Modellhöhe h_{Li} für das Liegende aus

$$h_{Li} \approx k \cdot \sigma_{Li} \tag{16}$$

wobei σ_{Li} die aus den entsprechenden Diagrammen nach Abb. 1 für den Stützpunktabstand l_i entnommenen Vorhersagefehler des Liegenden sind. Mit diesen Höhen können die Verluste und Verdünnungen nach der Beziehung (10) für das Hangende und nach

$$P_{Lai} = R_{Lai} = V_{Lai} = \frac{1}{3} \cdot \frac{1}{2} F \cdot h_{Li} \tag{17}$$

für das Liegende berechnet werden. Die Bezugsfläche F ergibt sich, indem von der Lagerstättenfläche F_E einer Abbaueinheit der Flächeninhalt F_P abgesetzt wird, den die Pfeiler darin einnehmen:

$$F = F_E - F_P \tag{18}$$

Bei Verschiebung der technologischen Abbaugrenze lassen sich die Verluste und Verdünnungen durch Interpolation zwischen den Fällen a, b und c ermitteln. Für die Fälle b und c gelten die Formeln (11) bis (14).

Ermittlung des Optimums durch Variantenvergleich

Der Lagerstättenvorrat Q_{FO} in der Bezugsfläche F wird über die Isolinien der Mächtigkeit der Basisvariante berechnet. Davon können bei Verwendung der Netzvariante i gewonnen werden:

$$Q_{Fi} = Q_{FO} - P_{Hi} - P_{Li} \tag{19}$$

Hinzu kommen die Verdünnungen

$$R_{Fi} = R_{Hi} + R_{Li} \tag{20}$$

Unter Berücksichtigung der mittleren Gehalte γ_Q und γ_R ergibt sich der Gehalt γ_{Q+R} für den in der Bezugsfläche ausgebrachten verdünnten Rohstoff $Q_{Fi} + R_{Fi}$ der Variante i zu

$$\gamma_{Q+R,i} = \frac{Q_{Fi} \cdot \gamma_Q + R_{Fi} \cdot \gamma_R}{Q_{Fi} + R_{Fi}} \tag{21}$$

Mit diesen Größen läßt sich berechnen, wie sich in der Variante i die spezifischen Gewinnungs- und Aufbereitungskosten sowie die Erkundungskosten gegenüber der Basisvariante verändern.

Durch die Verluste erhöhen sich die spezifischen Gewinnungskosten, weil bei nahezu gleichen Aufwendungen ein geringeres Finalprodukt anfällt. Durch die Verdünnungen entstehen im wesentlichen zusätzliche Kosten bei der Aufbereitung,

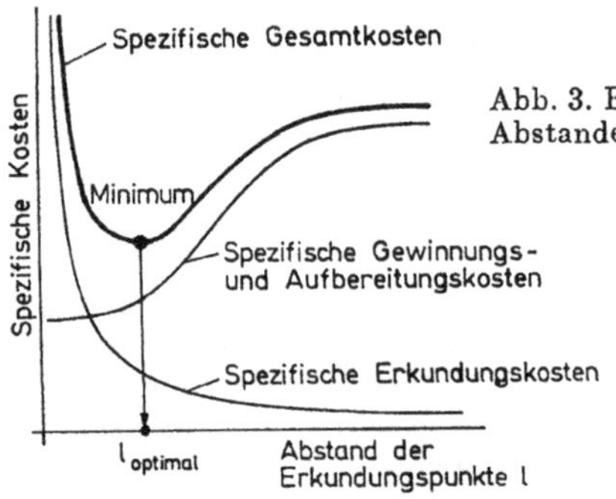

Abb. 3. Ermittlung des optimalen Abstandes der Erkundungspunkte

oder der Erlös reduziert sich wegen der verringerten Qualität. Die Senkung der spezifischen Erkundungskosten ergibt sich über die Anzahl der eingesparten Bohrungen.

Für die Auswahl des optimalen Erkundungsnetzes werden die spezifischen Erkundungskosten sowie die spezifischen Gewinnungs- und Aufbereitungskosten aller im Homogenitätsbereich untersuchten Netzvarianten in einem Diagramm nach Abb. 3 über den Stützpunktabständen l_i dargestellt (Menz 1983). Aus der Addition der spezifischen Erkundungs- sowie Gewinnungs- und Aufbereitungskosten (dünn ausgezogene Linien) ergeben sich die spezifischen Gesamtkosten (dick ausgezogene Linie). Das Minimum der spezifischen Gesamtkosten liefert den optimalen Abstand der Erkundungspunkte $l_{optimal}$ im untersuchten Homogenitätsbereich.

Bei Anisotropie genügt es, wenn beispielsweise die kurzen Seiten der Netzmaschen als Abszissen in Abb. 3 aufgetragen werden, weil das Verhältnis der Maschenweiten, das aus der Veränderlichkeit der bergbaugeologischen Kompliziertheit abgeleitet wird, konstant ist.

Zusammenfassung

In einem bereits erschlossenen, repräsentativen Gebiet einer Lagerstätte werden für verschiedene Netzvarianten die Verluste und Verdünnungen festgestellt. Es wird analysiert, wie sich von Variante zu Variante die spezifischen Erkundungs- sowie Gewinnungs- und Aufbereitungskosten ändern. Das Minimum dieser Kosten liefert die optimale Netzvariante, die unter der Berechtigung des Analogieschlusses auf die noch weiter zu erkundenden Gebiete übertragen wird. Die Untersuchung ist für alle Homogenitätsbereiche durchzuführen, in die die Lagerstätte auf Grund der bergbaugeologischen Kompliziertheit eingeteilt werden muß.

Резюме

На изученном и представительном участке одного из месторождений определялись при различных вариантах разведочной сетки объемы потерь и разубоживания. Проводится анализ изменения специфических затрат на разведку, добычу и обогащение, соответствующих различным вариантам сетки. Минимальные затраты соответствуют варианту оптимальной сетки, которая с учетом метода аналогии переносится на участки, где будут проводиться разведочные работы. Исследования необходимо проводить по всем однородным участкам, на которые делится месторождение, в соответствии с характером его горно-геологической сложности.

Summary

In an opened representative part of a deposit the losts and dilutions are established for some different net variants. The alteration of the specific prospection-, mining- and refining costs in each case is analyzed. The minimum of these costs determines the optimum net variant, which is transferred to the regions which have to be prospected if analogical deduction is permissible. The investigation has to carried out for all regions of homogenity, in which the deposit is classified because of mining-geological complexity.

Literatur

Bukrinskij, V. A.: Geometrija nedr. — Moskva: Izdat. Nedra 1985

Menz, J.: Die Bedeutung der Geostatistik für die Rationalisierung bergbaulicher Prozesse. — Neue Bergbautechnik, Leipzig 11 (1981) 11, S. 656–659

— Beitrag zur optimalen Erkundung. — Neue Bergbautechnik, Leipzig 13 (1983) 2, S. 73–76

— Lagerstättenmodellierung mit dem Programmsystem GEOSTATISTIK. — Neue Bergbautechnik, Leipzig 15 (1985) 11, S. 404–407

— Geostatistische Vorhersage mit Kleinrechnern und ihre Anwendung zur Profilkonstruktion. — Z. angew. Geol. Berlin 32 (1986) 6, S. 143–146

Ermittlung vorläufiger Konditionen bei der Gewinnung mineralischer Rohstoffe durch Laugung oder mikrobielle Prozesse

Horst Bachmann, Freiberg

Einleitung

Die Biotechnologie hat als eine der Schlüsseltechnologien bei der Intensivierung der Produktionsprozesse in der Land- und Nahrungsgüterwirtschaft, in der pharmazeutischen Industrie und in der Wasserwirtschaft verstärkt Eingang gefunden. Sie trägt dort in kostengünstiger Weise zur Erhöhung der Produktion und zur Erhaltung der Qualität der natürlichen Umwelt bei. Der Gedanke liegt nahe, auch für die Gewinnung und Verarbeitung mineralischer Rohstoffe die Anwendungsmöglichkeiten solcher Prozesse zu nutzen und neue, dafür geeignete Substanzen und Prozesse zu entwickeln. Die Bearbeitung solcher Vorhaben mit Blick auf eine mögliche industrielle Nutzung führt naturgemäß zu ökonomischen Fragen.

Eine Ausgangsüberlegung besteht darin, daß eine mikrobielle Rohstoffgewinnung oder Laugungsprozesse dann interessant werden, wenn herkömmliche Verfahren der mechanischen Anreicherung der Nutzkomponenten eines Rohhaufwerkes technologisch versagen oder zu untragbar hohen Kosten führen. Andererseits setzt eine mikrobielle oder chemische Stoffgewinnung bei einer hinreichend zuverlässigen Einschätzung der Kosten aller notwendigen Prozeßstufen die Einhaltung einer Mindestkonzentration der Nutzkomponente (Wertstoff) im Ausgangsstoff voraus. Deshalb muß ein analoger Konditionswert wie der industrielle Mindestgehalt bei der traditionellen bergbaulichen Rohstoffgewinnung festgelegt werden. Es ist zu ermitteln, welche Grenzkonzentration an Nutzkomponente unter Berücksichtigung der Verarbeitungsverluste nicht unterschritten werden darf.

Im folgenden wird hierzu ein methodischer Beitrag vorgestellt.

Eingang des Manuskripts in der Redaktion: 2. 6. 1987

Produktionsprozeß

Der Produktionsprozeß für eine mikrobielle Rohstoffgewinnung ist in mehrere Prozeßstufen zu gliedern, die sich nach Anzahl, Technologie und Kosten und für einzelne Rohstoffarten erheblich unterscheiden. Wenn als Endprodukt ein Metall oder eine Metallverbindung (Reinmetall oder Metalloxid) betrachtet wird und als Rohhaufwerk — hier als Input bezeichnet — ein Mineralgemisch unterschiedlicher chemischer und mineralogischer Zusammensetzung und physikalischer Beschaffenheit vorliegt, so können in der Reihenfolge vom Endprodukt zum jeweiligen Input folgende Prozeßstufen unterschieden werden:

1. In der letzten Prozeßstufe werden die mikrobiell oder durch Laugung gewonnenen, mit Wertstoff beladenen Biomassen oder angereicherten Lösungen zu einem der oben näher bezeichneten Endprodukte verarbeitet.
2. Als Input liegt ein fein- bis mitteldisperses Haufwerk vor. Auf einem speziell vorbereiteten Untergrund wird ähnlich wie bei der Haufenlaugung der mikrobielle Trenn- und Anreicherprozeß durchgeführt. Das erhaltene Substrat wird nach entsprechender Vorbehandlung auf chemischem oder metallurgischem Weg zum Endprodukt verarbeitet. Hier wie auch bei den folgenden Prozeßvarianten fällt ein unhaltiges Abfallprodukt an, das ordnungsgemäß deponiert werden muß, falls es nicht vollständig oder teilweise einem anderen Verwendungszweck zugeführt werden kann.
3. Bei Haldengestein (Bergbauhalden) mit mittlerer bis grober Körnung wird in der Regel eine Zerkleinerung auf eine für den Prozeß zweckmäßige Korngröße erforderlich sein. Auch hier ist eine Vorbereitung des Untergrundes erforderlich.
4. Wenn ein in der Erdkruste anstehendes festes oder rolliges Gestein als Input gewonnen und dann einer Verarbeitung zugeführt werden soll, werden außer den vielfältigen Vorbereitungsarbeiten geowissenschaftlicher Art auch Investitionen und Betriebskosten für die Tagebaugewinnung entstehen.
5. Soll ein Rohhaufwerk aus untertägig anstehenden Gesteinsmassen verarbeitet werden, sind drei grundsätzliche Möglichkeiten denkbar.

- Entweder erfolgt die mikrobielle Anreicherung durch Zuführung eines Reaktionsmediums über Bohrlöcher in vorhandene vorgerichtete Grubenräume, oder
- ebenfalls in situ nach einer speziellen bergmännischen Aus- und Vorrichtung in Magazinen oder Blöcken, in denen das Anstehende in Abhängigkeit von den mineralogischen, petrophysikalischen und hydrogeologischen Verhältnissen durch Auflockerungssprengungen für die Mineralbehandlung vorbereitet worden ist, oder
- es erfolgt eine traditionelle bergmännische Gewinnung und Förderung und die Weiterverarbeitung über Tage wie bei den anderen Prozeßstufen. Ebenso ist eine Kombination der üblichen bergmännischen Gewinnung mit der In-situ-Verarbeitung von Versatzmassen und Restvorräten möglich.

Prozeßkosten

Die Abschätzung der Prozeßkosten stellt bei der Konditionsermittlung den neuralgischen Punkt dar, sofern im konkreten Fall keine Ergebnisse von Durchführbarkeitsstudien vorliegen, die eine Vorkalkulation der Kosten erlauben.
Die folgenden Überlegungen gehen jeweils von den Kosten je Tonne Input aus. Je nach Anzahl der Prozeßstufen dürften sie den weiten Bereich von etwa einer Zehnerpotenz überdecken.
Während für die Prozeßstufen 3 bis 5 Erfahrungswerte vorliegen und Analogieschlüsse möglich sind und auch für die metallurgische oder chemische Endverarbeitung die Kosten abgeschätzt werden können, sind gegenwärtig für die mikrobielle Prozeßstufe Angaben kaum möglich.
Für eine allgemeine Methode muß von Prozeßkosten der jeweiligen Prozeßstufe, ausgedrückt als spezifische Kosten/t Input der ersten Prozeßstufe, ausgegangen werden.
Im Berechnungsbeispiel werden für die metallurgische oder chemische Verarbeitung eines schwach angereicherten Konzentrates folgende Erfahrungswerte angenommen:

- Wertstoffkonzentration des Inputs: 1000 g/t.
- Die Verarbeitung zu einem handelsüblichen Endprodukt bei einem Wertstoffausbringen von 0,8 erfordert für eine Tonne Endprodukt eine Menge von 1250 t Input in der Anfangsprozeßstufe (z. B. bergmännisches Rohhaufwerk).
- Die Verarbeitungskosten in der letzten Prozeßstufe sollen 6000,— M/t Endprodukt betragen.
- In diesem Falle wird jede Tonne Input mit 4,8 M belastet.

In Abb. 1 werden für verschiedene Prozeßvarianten die kumulativen Prozeßkosten in einem geschätzten Bereich von 10,— bis 200,— M/t Input angegeben. In diesen Angaben sind keine Folgekosten für Zwischenaufhaldung und Deponien berücksichtigt.
Für diese geschätzten Prozeßkosten wurde eine tägliche Leistung von 100 t Rohhaufwerk angenommen. Veränderte Durchsatzleistungen verursachen eine Kostendynamik. Höhere Leistungen führen zu einer Kostendegression. Aus verschiedenen Gründen ist zur Berücksichtigung der Kostendegression eine Aufteilung der Gesamtkosten in Investititionskosten und Betriebskosten (Gesamtkosten minus Abschreibungen) zweckmäßig. Für beide Kostenteile sind dann verschiedene Degressionsexponenten D anzusetzen.
Internationale Erfahrungen geben für die chemische Industrie und für den Bergbau Degressionsexponenten für Investitionen von 0,6 bis 0,8 an. Die Investitionskosten für veränderte Kapazitäten können dann nach folgender Beziehung ermittelt werden:

$$\frac{I_2}{I_1} = \frac{K_2^D}{K_1}$$

I_2 — Investitionen für eine Kapazitätsvariante K_2
I_1 — Investitionen bei einer Ausgangskapazität K_1
D — Degressionsexponent.

Für die Betriebskosten im Bergbau ist in der Regel der größere Degressionsexponent zu verwenden, der eine geringere Kostendegression bei größeren Kapazitäten ausdrückt. Im Falle der mikrobiellen Verarbeitung ist die Prozeßdauer von erheblichem Einfluß auf die Betriebskosten (hier vor allem auf die Materialkosten). Wenn man davon ausgeht, daß sich die Betriebskosten angenähert proportional zu der Prozeßdauer verhalten, kann vermutlich von einer Kostendegression der Betriebskosten bei veränderten Kapazitäten abgesehen werden.

Konditionsermittlung

Zur Ermittlung einer Mindestkonzentration an Nutzkomponenten müssen folgende Werte bekannt sein oder mit hinreichender Genauigkeit abgeschätzt werden können:

- die Stufenkosten K_i für eine Tonne Ausgangsprodukt;
- der Preis für eine Tonne Endprodukt. Im Regelfall ist vom Preis der Gewinn je Tonne Endprodukt abzuziehen. Aus Gründen der Vereinfachung wird im Modellfall darauf verzichtet.
  In den Fällen, in denen der volkswirtschaftlich vertretbare Aufwand für das Endprodukt höher als der Preis ist, tritt diese Größe (vvA) an Stelle des Preises.
- das Ausbringen an Nutzkomponente im Endprodukt, bezogen auf eine Tonne Ausgangsprodukt. Das Ausbringen ist $r < 1$.

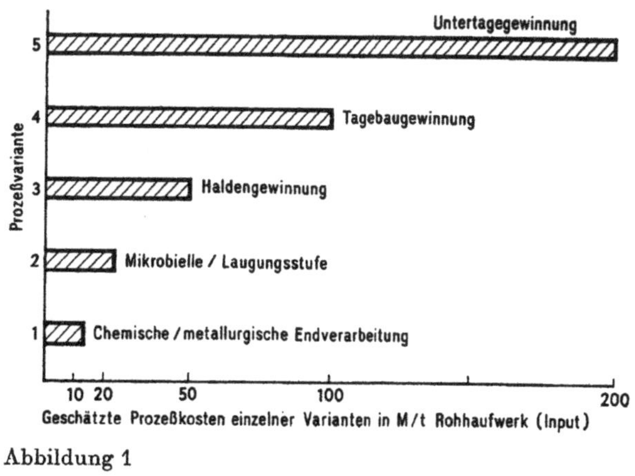

Abbildung 1

Ausgangsgleichung ist eine Kosten-Erlösfunktion, die den gesuchten Konditionswert als Unbekannte enthält:

$$\sum K_i = 10^{-6}\, c_m p r \tag{1}$$

In dieser Gleichung sind die Stufenkosten zur Verarbeitung von einer Tonne Input dem Erlös gegenübergestellt, der aus dem Endprodukt erzielt werden kann, das bei der Verarbeitung einer Tonne Input (Rohhaufwerk) anfällt.

K_i — Kosten der Prozeßstufe i in M/t Ausgangspunkt
c_m — Mindestkonzentration an Nutzkomponente im Input in g/t Rohhaufwerk
p — Preis des Endproduktes in M/t
r — Ausbringen an Nutzkomponente; $0 < r < 1$.

Wenn in (1) drei von vier Größen bekannt sind, lassen sich durch Umstellung folgende Fragen beantworten:

a) Wie hoch müßte der volkswirtschaftlich vertretbare Aufwand sein, wenn ein Ausgangsprodukt mit einer Wertstoffkonzentration von c vorliegt, das Ausbringen eingeschätzt werden kann und die Stufenkosten vorkalkuliert werden können?

$$vvA = \frac{10^6 \sum K_i}{cr} \tag{2}$$

b) Welches Ausbringen an Nutzkomponente muß erzielt werden, wenn bei einer festgestellten Konzentration an Nutzkomponente der Gewinnungsprozeß bei einem festen Preis erfolgreich durchgeführt werden soll und die Stufenkosten vorkalkuliert werden können?

$$r = \frac{10^6 \sum K_i}{cp} \tag{3}$$

c) Wie hoch muß die Konzentration an Nutzkomponenten im Input sein (Mindestkonzentration), wenn bei festem Preis p

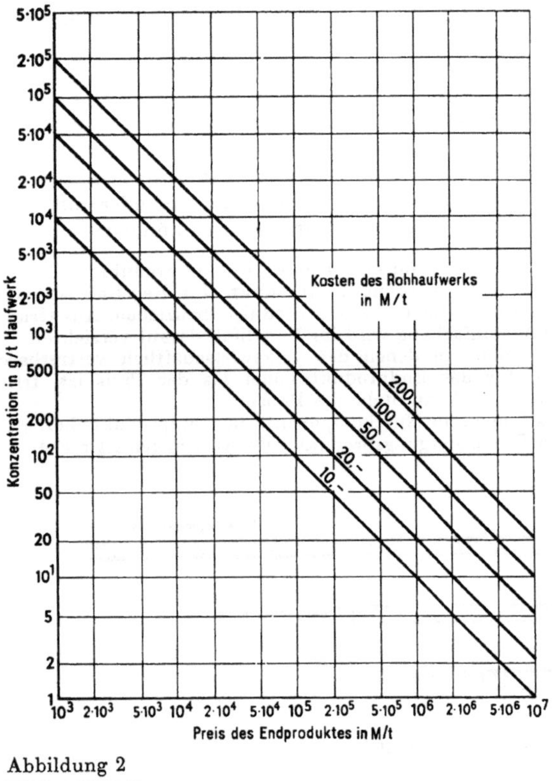

Abbildung 2

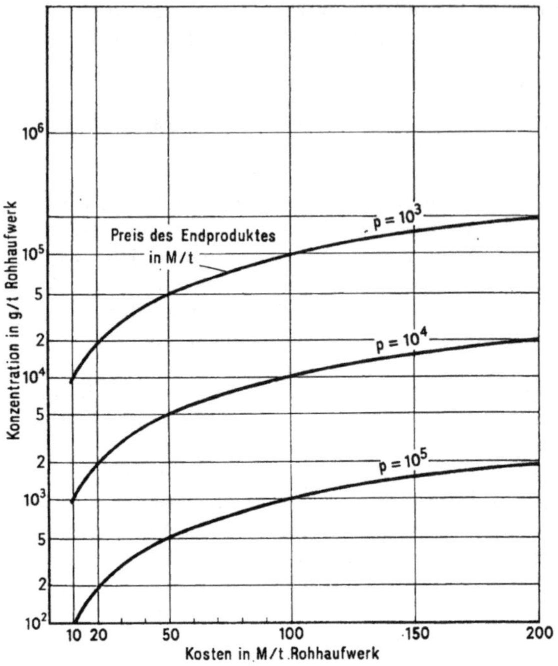

Abbildung 3

Werte für Stufenkosten und Ausbringen angegeben werden können?

$$c_m = \frac{10^6 \sum K_i}{pr} \tag{4}$$

Für Variantenuntersuchungen können Diagramme genutzt werden, deren Koordinaten doppelt oder einfach logarithmisch angelegt sind (Abb. 2 u. 3). Dadurch lassen sich recht große Konzentrationsspannen leicht überblicken.[1])

Wenn es sich um einen komplexen Ausgangsstoff handelt, der mehrere Nutzkomponenten j mit den Konzentrationen c_j bei möglichen Ausbringenskoeffizienten r_j enthält, ist eine Konditionsermittlung mit Hilfe des Diagramms nicht möglich. Die Fragestellung muß in diesem Fall lauten:

Welche Kosten je Tonne Input dürfen aufgewandt werden, damit die Summe der Erlöse aus den einzelnen Endprodukten nicht überschritten wird?

Die Berechnung muß dann über die Preisäquivalenz aus dem gewogenen arithmetischen Mittel der Erlöse und den Produkten aus Konzentration und Ausbringenskoeffizienten der einzelnen Komponenten erfolgen:

$$\sum K_i = \frac{10^6 \sum p_j c_j r_j}{\sum c_j r_j} \tag{5}$$

Wenn das Ziel des Prozesses nicht in erster Linie die Rohstoffgewinnung ist, sondern die Beseitigung von Schadstoffen aus Abwässern oder Altstoffen, sind methodisch zwei Wege denkbar.

1. Ist das Ziel ausschließlich die Beseitigung von Schadstoffen, ohne daß ein verwertbares Endprodukt zu gewinnen ist, tritt an die Stelle des Preises für ein Endprodukt die zu vermeidende Schädigung der Umwelt. Sie kann entweder im Geldausdruck als vermiedene Sanierungskosten ausgedrückt werden oder ist durch Experten bei monetär nicht erfaßbaren Faktoren zu beurteilen.
2. Wenn außer der Schadstoffbeseitigung gleichzeitig eine verwertbare Komponente gewonnen werden kann, ist der Preis dieser Nutzkomponente um die vermeidbaren Folgekosten anteilig zu erhöhen.

[1]) Zur Arbeit mit solchen c/p/K-Diagrammen ist eine großformatige Darstellung (A 2 oder A 3) zu empfehlen.

Mathematisch-statistische Ableitung von Fehlerparametern für kohlechemische Analysendaten innerhalb der Erkundung fester Brennstoffe

GÜNTER KAHNT, Altenburg

Einleitung

Mit der Herausbildung zentraler Laboratorien mit einer hochentwickelten Meßtechnik sind günstige Bedingungen für die Einführung der statistischen Qualitätskontrolle entstanden. Analysiert man das Aufgabenvolumen eines Zentrallaboratoriums der Braunkohleindustrie, so können die Hauptaufgaben in folgende Gruppen eingeteilt werden:

1. Analysen in großer Stückzahl auf gleichzeitig mehrere Analysenparameter von Proben bekannter Herkunft:
   - kohlechemische Untersuchungen von Erkundungsbohrungen aus bereits vorerkundeten Kohlefeldern,
   - laufende Qualitätsüberwachung der Produktion in Tagebauen, Brikettfabriken und Kraftwerken;
2. Analysen in großer Stückzahl auf gleichzeitig mehrere Analysenparameter von Proben unbekannter Herkunft:
   - kohlechemische Untersuchungen von Erkundungsbohrungen neuer Kohlefelder,
   - Proben aus Produktionsanlagen mit einer neuentwickelten Technologie;
3. Analysen in kleiner Stückzahl auf gleichzeitig mehrere Analysenparameter von Proben bekannter oder unbekannter Herkunft:
   - Reklamationsanalysen, Untersuchung von Spezialbrennstoffen, Ablagerungs- und Korrosionsprodukten u. a.

Für Analysen, die in einer großen Stückzahl serienmäßig durchgeführt werden, lohnt es, die statistische Qualitätskontrolle nach Schaffung der notwendigen Voraussetzungen einzuführen. Die meisten Analysenverfahren fordern Doppel- bzw. Dreifachbestimmungen. Trotz der damit mindestens hundertprozentigen Kontrolle der Analysenwerte schützt diese sehr aufwendige Arbeitsweise nicht vor Fehlanalysen bzw. großen Zufallsfehlern, weil durch diese Arbeitsweise nicht exakt auf die Richtigkeit und Zuverlässigkeit der Analysenergebnisse geschlossen werden kann (KAHNT 1972, 1975). Letztlich gestattet nur ein objektives und wirksames Kontrollprogramm, ein optimales Verhältnis zwischen Qualität, Quantität und Ökonomie im Laboratorium praktisch zu erreichen.

Auswertungsgrundlagen

Für die Qualitätsüberwachung und -regulierung von Serienanalysen sind Kontrollkarten ein ausgezeichnetes Hilfsmittel. Durch sie wird es möglich, vorbeugend und zielgerichtet in die laufenden Prozesse einzugreifen. Die Arbeitsgänge für die Auswahl der Kontrollkartenart, die Berechnung der einzelnen Kontrollkartenparameter und ihre Aussage wurden vielfach in der Literatur erörtert (KAHNT 1975). Neben den allgemeinen Grundsätzen, die sich aus den Kontrollkartenbildern für einen bestimmten zu überwachenden Prozeß ableiten lassen, die der Analytiker kennen und anwenden muß, wirken in der Praxis noch andere Faktoren. Für die Laborpraxis haben sich folgende Kontrollkarten bewährt (KAHNT 1975):

- x — Einzelwertkarte,
- D — Kontrollkarte (D = zulässige Differenz nach PEARSON) für Doppelbestimmungen,
- s — Karte (s = Standardabweichung),
- $\bar{x}$ — Mittelwertkarte ($\bar{x}$ = Mittelwert).

Aus einem genügend großen Analysenvorlauf ($n \geqq 30$) werden die einzelnen Parameter für einfache oder kombinierte Kontrollkarten berechnet (KAHNT 1975; STORM 1965). Die Warn- und Kontrollinien sind für bestimmte statistische Sicherheiten festzulegen (z. B. Warnlinien = 95%, Kontrollinien = 99%). Folgende Aussagen lassen sich u. a. aus dem Kontrollkartenbild ableiten:

Eingang des Manuskripts in der Redaktion: 2. 6. 1987

1. Streuen die Kontrollergebnisse regellos zwischen der oberen und der unteren Warnlinie der Kontrollkarte, so ist der Prozeß stabil und das Analysenverfahren liefert richtige Analysenwerte.
2. Zeigen die Kontrollergebnisse einen bestimmten „Gang", eine „Periodizität" oder nähern sie sich systematisch einer Warnlinie, so ist das Analysenverfahren instabil. Alle Analysenparameter sind einer systematischen Kontrolle zu unterziehen, um die vielfältig möglichen Fehlerursachen feststellen und beseitigen zu können. Erst bei einer nachgewiesenen Prozeßstabilität darf das Kontrollsystem wieder angewendet werden.
3. Überschreiten die Kontrollergebnisse erst mehrfach die Warnlinie und danach die Kontrollinie, so ist der Prozeß außer Kontrolle und die systematische Kontrolle hat zu erfolgen.

Besonders aus kombinierten Kontrollkarten für Einzelwerte, Mittelwerte, Standardabweichungen oder zulässige Differenzen nach PEARSON für gleichzeitig zu untersuchende Proben und Standardproben läßt sich auf einfache Weise eine Vielzahl von Informationen ableiten (KAHNT 1980; DOERFFEL, MÜLLER & UHLMANN 1986).

Die Korrelations- und Regressionsanalyse erlauben die Aufdeckung und quantitative Charakterisierung von Zusammenhängen zwischen verschiedenen Meßgrößen (KAHNT 1975; DOERFFEL 1982). Die anzuwendenden Formeln für die vorwiegend vorkommenden linearen und einfachen Zusammenhänge ungruppierter und quantitativ erfaßbarer Wertepaare sind u. a. in KAHNT (1975) enthalten.

Inhalt und Aufbau des Kontrollverfahrens

Ausgehend von dem zu lösenden Problem — Sicherung von richtigen und reproduzierbaren Analysenwerten innerhalb größerer Analysenserien — wird das Kontrollsystem in drei Teile gegliedert:

- laufend regulierende Kontrolle des Analytikers,
- auswertende Stichprobenkontrolle und
- gezielte korrelative Nachkontrolle.

Für die ersten zwei Teile des dreistufigen Kontrollsystems werden Kontrollkarten verwendet, deren Aufbau und Deutung klar und einfach sind. Die laufend regulierende Kontrolle besteht dabei aus zwei parallelen Kontrollkarten. Einerseits erfolgt die Analysierung von Standmusterproben, deren Qualitätsparameter bekannt sind und die in ihrer Probenzusammensetzung den Proben der laufenden Analysenserie entsprechen, und andererseits werden von den zu bearbeitenden Analysen Doppelbestimmungen durchgeführt. Die Ergebnisse der Standmusteruntersuchungen werden in einer Einzelwertkontrollkarte und die Differenzen zwischen je zwei aufeinander folgenden Analysen nach ihrem Betrag und ihrem Vorzeichen (Erstanalysenwert — Kontrollanalysenwert) in der D-Kontrollkarte erfaßt.

Die Differenzen der Doppelbestimmungen werden ebenfalls in die D-Kontrollkarten, besonders gekennzeichnet entsprechend ihrer Zusammensetzung nach Vorzeichen und Betrag, eingetragen. Aus diesen kombinierten Kontrollkarten lassen sich folgende Aussagen ableiten:

1. Streuen die Standmustereinzelergebnisse regellos zwischen den zwei Warnlinien der Einzelwertkontrollkarte, so ist der Prozeß stabil, und das Analysenverfahren liefert richtige Analysenwerte.
2. Streuen die Differenzen zwischen je zwei Standmusterwerten und der Doppelbestimmungen zufällig in ihrem Betrag und in ihrem Vorzeichen zwischen den Warnlinien, so liefert das Analysenverfahren reproduzierbare Analysenwerte.
3. Treten zwischen den Differenzen der Standmusterproben und der Doppelbestimmungen größere Abweichungen in ihrem Betrag auf, so muß das Analysenverfahren näher untersucht werden. Größere Differenzen bei den Doppelbestimmungen werden meist durch eine unzureichende Probenvorbereitung verursacht. Durchweg kleinere Differenzen bei den Doppelbestimmungen verlangen die kritische Nachprüfung der vorgegebenen Kontrollgrenzen. Die Verknüpfung dieses Differenzenvergleiches hat den Vorteil, daß gleichzeitig vorangegangene Arbeitsschritte mit in der Kontrolle erfaßt werden, und es zwingt außerdem zur dynamischen Arbeitsweise mit diesen Kontrollkarten. Die richtige Beherrschung beider Prozesse erlaubt, ein Höchstmaß an Qualitätssicherung zu garantieren.
4. Zeigen die Kontrollparameter einen Gang in der Kontrollkarte bzw. überschreiten sie die Warnlinien, so sind sofort das Analysenverfahren und die Arbeitsweise des Analytikers zu überprüfen. Die Stichprobenkontrolle wird bis zum Erreichen der Prozeßstabilität verstärkt. Überschreiten die Kontrollparameter die Kontrollinien, so setzt die hundertprozentige Kontrolle ein.

Die Kontrollkarten für die regulierende Kontrolle werden vom Analytiker unmittelbar am Arbeitsplatz geführt und ausgewertet. Er wird damit befähigt, die Qualität seiner Arbeit konkret abrechenbar zu beurteilen und außerdem sofort vorbeugende Maßnahmen zu treffen, wenn sich aus der Tendenz der Kontrollparameter Unregelmäßigkeiten abzeichnen, die eine baldige Prozeßinstabilität bewirken können.

Kann mit dem ersten Kontrollschritt für eine größere Analysenserie (50 bis 200) mathematisch-statistisch die Prozeßstabilität nachgewiesen werden, so wird der geprüfte Posten von Analysenwerten der auswertenden Stichprobenkontrolle unterzogen. Vom zuständigen Leiter wird aus dem Gesamtposten eine bestimmte Anzahl von Stichproben zufällig entnommen und dem Analytiker verschlüsselt zur wiederholten Analysierung übergeben. Die erzielten Differenzen erfaßt man wieder nach ihrem Betrag und ihrem Vorzeichen in einer D-Kontrollkarte, die zweckmäßigerweise gleichzeitig für mehrere Gehaltsspannen benutzt werden kann. Wenn für die Warn- und Kontrollinien die jeweiligen speziellen Untersuchungsbedingungen (Wiederhol- bzw. Vergleichsstreubereich) exakt beachtet werden, gestatten diese Kontrollkarten einen objektiven Vergleich mit den D-Kontrollkarten der laufend regulierenden Kontrolle des Analytikers. Gefälschte Kontrollkarten werden sichtbar, und zeitlich bedingte Veränderungen (Raumtemperaturverhältnisse bei kalorimetrischen Bestimmungen, Veränderungen des Probegutes, Verschleiß von Anlagenteilen, veränderte Reaktionsbedingungen u. a.) müssen bei der Auswertung Beachtung finden. Überschreiten die Kontrollwerte die Warn- bzw. Kontrollinien, so wird wie bei der laufenden Kontrolle verfahren.

Wird eine Vielzahl von Proben gleichzeitig durch mehrere Qualitätsmerkmale charakterisiert, die untereinander korreliert sind, so können die entsprechenden Wertepaare durch eine gezielte korrelative Nachkontrolle überprüft werden. Aus einem genügend großen Analysenvorlauf, früheren gleichgearteten Messungen bzw. technologisch bedingten Zusammenhängen erfolgt die Berechnung der erforderlichen Daten. Ist der Korrelationskoeffizient eines logischen Zusammenhanges statistisch gesichert, so werden die Koeffizienten der Regressionsgeraden und die dazugehörigen Varianzen berechnet. Je enger der korrelative Zusammenhang ist, desto kleiner sind die Varianzen und somit auch die Kontrollgrenzen für die korrelative Nachkontrolle der Analysenwerte. Da die Varianzen bei heterogenen und kompliziert zusammengesetzten Untersuchungsmaterialien relativ groß sind, wird man bereits die $\pm 1{,}96 \cdot s$-Grenzen als Kontrollinien verwenden, um eine ausreichende Aussagekraft zu erzielen. Dieser Kontrolle werden sämtliche Analysenwerte unterzogen. Entsprechen die Meßwerte nicht dem vorher berechneten Zusammenhang, dann müssen sie einer zusätzlichen Nachkontrolle unterworfen werden. Gleichzeitig sind bei größeren und systematischen Abweichungen der Korrelationskoeffizient und die anderen Berechnungsgrößen mit neuen Meßreihen mathematisch-statistisch zu testen und eventuell zu verändern. Bei der Nachkontrolle der Analysenwerte empfiehlt es sich, stets eine neue Analysenprobe aus der Ausgangsprobe herzustellen, da die Ausreißerwerte oft aus Probenvorbereitungsfehlern resultieren.

Für die einzelnen Kontrollschritte haben sich nach den bisher vorliegenden praktischen Erfahrungen folgende Kontrolldichten (Prozent der Gesamtanalysenanzahl) bewährt, die natürlich sehr stark vom Probegut, der Stabilität des Analysenverfahrens und der Arbeitsweise des Analytikers abhängen:

— laufende Standmusterkontrolle = 10%
— laufende Doppelbestimmungen = 15%
— auswertende Stichprobenkontrolle = 15%
— gezielte korrelative Nachkontrolle = 10%.

Zur Einführung des beschriebenen Kontrollverfahrens für chemische Serienanalysen sind folgende Bearbeitungsstufen notwendig:

1. Untersuchung der Analysenverfahren zur Erbringung des Nachweises, daß die analytisch-chemischen Prozesse stabil ablaufen. Berechnung der erforderlichen mathematisch-statistischen Fehlerparameter entsprechend den praktizierten Untersuchungsbedingungen;
2. Auswahl von Standmusterproben, die der Gesamtheit der durchzuführenden Analysen weitestgehend entsprechen, Berechnung der Fehlerparameter aus dem Analysenvorlauf und Vergleich mit früher abgeleiteten Werten;
3. Berechnung der Kontroll- und Warnlinien, Festlegung der Kontrolldichte und Entwicklung einer Kontrollkarte;
4. Ermittlung logischer korrelativer Zusammenhänge und Berechnung der erforderlichen Daten für die gezielte Nachkontrolle;
5. Festlegung von Zusatzbedingungen für die Wahrung der zufälligen Stichprobenentnahme und Beachtung von Faktoren, die zeitliche Veränderungen beispielsweise des Probegutes hervorrufen können (Wassergehalt, Teerausbeute u. a.).

Nur die konsequente Einhaltung aller Verfahrensschritte gewährleistet die erfolgreiche Einführung des mathematisch-statistischen Kontrollsystems.

Kohlechemische Untersuchung eines Erkundungsobjektes als Anwendungsbeispiel

Bei der Erkundung von Lagerstätten fester Brennstoffe ist die Kenntnis der chemischen Zusammensetzung der anstehenden Kohleflöze für die qualitätsmäßige Bilanzierung der Kohlearten von wesentlicher Bedeutung. Die Bohrproben werden dabei in der Regel nach dem bei KAHNT (1983) dargestellten Schema aufbereitet und analysiert. Obwohl bereits von den Geologen ein umfangreiches Kontrollsystem geschaffen wurde, bringt die Anwendung des vorgeschlagenen komplexen Kontrollverfahrens eine wesentliche Erhöhung der Aussagekraft.

Von mindestens 100 gemeinsam mit dem Erkundungsgeologen auszuwählenden Proben, die das zu erkundende Kohlefeld repräsentieren, werden im Analysenvorlauf die wichtigsten Analysenparameter bestimmt. Daraus wählt man entsprechend der Häufigkeitscharakteristik etwa zehn Standmusterproben für den gesamten Analysenumfang aus (KAHNT 1975). Die Kontrollparameter werden entweder aus früher abgeleiteten und gesicherten Werten oder aus speziellen Versuchsreihen berechnet. Die Grundlagen für die zwei ersten Kontrollschritte sind somit geschaffen. Kann innerhalb der sich anschließenden Serienanalysen bei der laufenden Standmusterkontrolle und der auswertenden Stichprobenkontrolle der Nachweis erbracht werden, daß ein größerer Posten von Analysenwerten aus stabilen analytisch-chemischen Prozessen resultiert, so folgt abschließend die korrelative Nachkontrolle sämtlicher Wertepaare. Die Minimierung der Schwankungsbreite der berechneten Werte gelingt oft durch Gruppenbildung bzw. Einbeziehung eines dritten Parameters. Bei der Untersuchung fester Brennstoffe kann eine Vielzahl von Analysenparametern für derartige Berechnungen ausgenutzt werden. Einmal berechnete korrelative Zusammenhänge dürfen jedoch niemals mechanisch auf andere Kohlequalitäten übertragen werden. In jedem Falle sind spezielle Untersuchungen durchzuführen und kritisch zu bewerten.

In diesem Zusammenhang soll auch erwähnt werden, daß korrelativ berechnete Werte mit Vorteil zur Analysenkontrolle ausgenutzt werden können, aber die experimentellen Werte nicht vollkommen ersetzen. Für berechnete Regreßwerte sind stets die dazugehörigen Vertrauensintervalle bei der Anwendung zu berücksichtigen, um ein Höchstmaß an Aussage abzusichern (KAHNT 1980). Das beschriebene dreistufige Kontrollsystem läßt sich auch bei der laufenden Überwachung von Produktionsanlagen und anderen Aufgaben vorteilhaft anwenden. Neben einer erhöhten Qualitätssicherung wird eine Steigerung der Arbeitsproduktivität bei gleichzeitiger Kostensenkung gegenüber der Arbeitsweise mit pauschalen Doppelbestimmungen möglich. Die konsequente Anwendung dieses Kontrollsystems erlaubt ebenfalls eine noch objektivere materielle Stimulierung der Arbeitsleistungen der Werktätigen (KAHNT 1975).

Mathematisch-statistische Ableitung von Fehlerparametern

Mathematisch-statistische Fehlerparameter

Die Standardabweichung s ist ein objektives Maß für die Leistungsfähigkeit eines Analysenverfahrens. Die Berechnung der Standardabweichung ist einfach und kann nach mehreren Varianten erfolgen (KAHNT 1975). Mit ihr lassen sich die maximal zulässige Differenz (D) zwischen einer verschiedenen Anzahl von Einzelwerten und das Vertrauensintervall des jeweiligen Mittelwertes ($\Delta\bar{x}$) berechnen. Als statistische Sicherheit wurde 95% festgelegt.

Die Versuchsbedingungen werden wie folgt definiert:

Wiederholstreubedingungen:

eine Probe, ein Gerät, ein bestimmtes Analysenverfahren, ein Analytiker

Vergleichstreubereich:

eine Probe, Geräte gleicher Bauart, ein bestimmtes Analysenverfahren, verschiedene Analytiker

Die wichtigsten Berechnungsformeln sind:

D = zulässige Differenz nach PEARSON (KAHNT 1975):

$$D = x_{max} - x_{min} < D(P, nj) \cdot s$$

x_{max} — maximaler Einzelwert
x_{min} — minimaler Einzelwert
P — mathematisch-statistische Sicherheit
s — Standardabweichung
nj — Anzahl der Einzelbestimmungen
P — PEARSONscher Faktor 2,77 bei nj = 2
3,31 bei nj = 3

$\Delta\bar{x}$ = Vertrauensintervall des Mittelwertes (KAHNT 1975):

$$\Delta\bar{x} = \frac{t(P, f) \cdot s}{\sqrt{nj}}$$

P — mathematisch-statistische Sicherheit
f — Anzahl der Freiheitsgrade
t(P, f) — Integralgrenze der t-Verteilung
nj — Anzahl der Einzelbestimmungen

Mit diesen einfachen mathematisch-statistischen Berechnungsmethoden lassen sich eindeutige und objektive Fehlerparameter ableiten.

Vorschlag von Fehlerparametern für einige Hauptanalysenparameter

In den Tab. 1 bis 3 sind einige mathematisch-statistisch abgeleitete Fehlerparameter zusammengestellt, die es zu prüfen und zu präzisieren gilt, um besonders für geologische Zwecke (innere und äußere Analysenkontrolle) endgültig repräsentative Angaben zu erhalten.

Tab. 1. Vorschlag über zulässige Fehlerparameter

Analysenverfahren	Wiederholstreubereich			Vergleichstreubereich			gB¹)
	TGL	s	D (nj = 2)	TGL	s	D (nj = 2)	
W_t^r/Wa, in %							1
≦ 20	0,3	wie TGL bzw.		0,5	wie TGL bzw.		
> 20	0,6 max.	kleiner		0,8 max.	kleiner		
A^d, in %							2
≦ 10	0,3	0,2	0,5	0,4	0,3	0,5	
> 10	0,7 max.	0,3	0,8	1,0 max.	0,5	1,4	
Q_s^a/Q_s^d, in J/g	120	72	200	330	150	420	3
$S_t^{a/d}$, in %							
≦ 2,0	0,05			0,10			
> 2,0–6,0	0,10	0,10	0,30	0,20	0,20	0,6	
> 6,0	0,15			0,30			
$S_c^{a/d}$, in %							4
≦ 2	0,10			0,15			
> 2	0,20	0,08	0,20	0,30	0,15	0,40	
B^d, in %							5
≦ 5	0,3			0,5			
≧ 5–10	0,5	0,20	0,5	0,7	0,30	0,80	
≧ 10–15	0,7			0,9			
> 15	1,0			1,4			
$Teer^d$, in %	±0,25	0,35	1,0	±0,35	0,54	1,50	6
$Koks^d$, in %	±0,35	0,55	1,5	±0,50	0,80	2,20	6

¹) siehe Abschnitt „Grundsätzliche Bemerkungen....“

Tab. 2.

Analysenverfahren	Wiederholstreubereich			Vergleichstreubereich			gB
	TGL	s	D (nj = 2)	TGL	s	D (nj = 2)	
$C^{a/d}$, in %	±0,15	0,50	1,38	±0,25	0,70	2,00	7
$H^{a/d}$, in %	±0,10	0,15	0,40	±0,15	0,20	0,60	7
U^d, in %							
≦ 5	±0,10	0,10	0,30	±0,15	0,15	0,40	
> 5	±0,20	0,20	0,60	±0,30	0,30	0,80	
$Z^{r/d}$, in %	±0,10	0,10	0,30	±0,20	0,20	0,60	8
$Cl^{a/d}$, in %	10% rel.	ist noch zu bestimmen		15% rel.	ist noch zu bestimmen		
Na_2O^d, in %							9
≦ 0,5	±0,03			±0,05			
> 0,5	±0,06	0,02	0,05	±0,15	0,04	0,10	
K_2O^d, in %							9
≦ 1,0	±0,05			±0,10			
> 1,0	±0,10	0,02	0,05	±0,15	0,04	0,10	
ASV, in °C							10
t_A	±10	10	30	–	15	40	
t_B	±10	7	20	–	10	30	
t_C	±15	7	20	–	10	30	

Tab. 3.

Ascheanalyse	ca. bzw. >	Wiederholstreubereich			Vergleichstreubedingungen			gB
		TGL	s Vorschlag	D (nj = 2)	TGL	s Vorschlag	D (nj = 2)	
% SiO_2	20	±0,25	0,50	1,40	±0,50	1,0	2,5	11
% Fe_2O_3	20	±0,25	0,40	1,10	±0,50	1,0	2,5	
% Al_2O_3	20	±0,25	0,50	1,40	±0,50	1,0	2,5	
% CaO	20	±0,25	0,30	0,80	±0,50	1,0	2,5	
% MgO	10	±0,20	0,30	0,80	±0,30	0,6	1,5	
% SO_3	10	±0,15	0,30	0,80	±0,25	0,6	1,5	
% TiO_2	1	±0,10	0,05	0,14	±0,20	0,1	0,3	

s — Standardabweichung
D (nj = 2) — zulässige Differenz nach PEARSON zwischen 2 Einzelbestimmungen

Grundsätzliche Bemerkungen zu den vorgeschlagenen Fehlerparametern

1. Angaben sind nur für stückige Braunkohle gültig. Probleme der Oberflächenfeuchte sind zusätzlich zu betrachten.
2. Die Fehler sind in Abhängigkeit vom absoluten Aschegehalt exakt zu untersuchen, ebenfalls die konkreten Verhältnisse bei salzhaltigen Brennstoffen.
3. Die Fehlerfortpflanzung wirkt enorm. Es gelten u. a. folgende Faktoren

$$Q_s{}^a = 1 \quad Q_s{}^d = 2{,}3 \quad Q_s{}^{daf} = 3{,}9 \quad Q_s{}^r = 2{,}3$$

$$Q_i{}^a = 1{,}8 \quad Q_i{}^d = 3{,}3 \quad Q_i{}^{daf} = 5{,}2 \quad Q_i{}^r = 2{,}8.$$

4. S_c nach TGL 42851 (Entwurf 1984) ist in seiner technischen Umsetzung fraglich (Verbrennungstemperatur, -art).
5. Die Bestimmung des Benzenlöslichen ist technisch fast bedeutungslos. Mathematisch-statistisch gesicherte Werte für das Toluenlösliche liegen noch nicht vor.
6. Keine der Apparaturen ist handelsüblich, deshalb vermutlich die zu hohen Zufallsfehler.
7. In der Regel arbeitet man im Halbmikromaßstab, so daß die hohen Zufallsfehler bis hin zur „Ausreißerhäufigkeit" bei heterogenen Brennstoffproben gegeben sind.
8. Die zulässigen Differenzen sind für die Bestimmung des Z-Gehaltes noch zu differenzieren:

$$< 1\%; \quad 1\text{–}5\%; \quad > 5\%.$$

9. Definition „Salzkohle" ist neu und umfassender festzulegen.
10. Keines der Geräte ist seit Jahren mehr handelsüblich. Die dadurch notwendigen „Eigenentwicklungen" bedingen in der Regel größere Fehler und Unsicherheiten.
11. Die praktisch erreichbaren und nach TGL geforderten Fehlerparameter weichen stark voneinander ab. Zur Klärung dieses Sachverhaltes sind gezielte Grundsatzuntersuchungen notwendig.

Zusammenfassung

Inhalt, Ziel und Anwendungsmöglichkeiten eines dreistufigen komplexen Kontrollsystems für Serienanalysen auf der Grundlage der Methoden der mathematisch-statistischen Qualitätskontrolle werden erörtert. Mit diesem Kontrollsystem wurde eine Vielzahl kohlechemischer Analysenverfahren und -daten kritisch untersucht. Die daraus abgeleiteten quantitativen Fehlerparameter werden angegeben und diskutiert. Es wird vorgeschlagen, ein derartiges Kontrollsystem auch bei der „inneren" und „äußeren" Kontrolle von kohlechemischen Erkundungsdaten anzuwenden, um die Repräsentanz des gesamten Analysenmaterials weiter zu verbessern.

Резюме

Рассматривается содержание, цели и возможности использования трехступенчатой комплексной контрольной системы серийных анализов на основе методов математическо-статистических проверок качества. С помощью этой контрольной системы было проведено большое количество углехимических методов анализов и их данных. Представлены и обсуждаются полученные количественные параметры ошибок. Предлагается использовать такую контрольную систему также и для „внутреннего" и „внешнего" контроля углехимических данных разведки с целью дальнейшего повышения представительности всего аналитического материала.

Summary

Content, mark and possibilities of use of a three-stage complex control system for analyses of seria on the base of methods of the mathematic-statistical quality control are discussed. With the help of this control system a number of coal-chemical analyses and data were investigated. The deduced quantitative parameters of error are given and discussed. It is proposed to use such a control system also in the "inner" and "upper' control of coal-chemical data of prospection to improve the representativeness of the whole material of analyses.

Literatur

DOERFFEL, K.: Statistik in der analytischen Chemie. — Leipzig: VEB Deutscher Verlag für Grundstoffindustrie 1982
DOERFFEL, K.; MÜLLER, H.; UHLMANN, M.: Prozeßanalytik. — Leipzig: VEB Deutscher Verlag für Grundstoffindustrie 1986
KAHNT, G.: Methoden zur Beurteilung von Analysenverfahren und Analysenwerten. — IfK-Informationen, Dresden 60 (1972) S. 60—65
— Kritische Untersuchungen einiger Methoden zur Analysierung fester Brennstoffe und Erarbeitung eines Kontrollsystems für Serienanalysen. — Freiberg: Bergakademie Freiberg, Diss. 1975 (frei für Lit.-Tausch)
— Komplexes Kontrollsystem für kohlechemische Serienanalysen innerhalb der laufenden Qualitätskontrolle und der Erkundung fester Brennstoffe.— Neue Bergbautechnik, Leipzig 10 (1980) 8, S. 455—460
— Probleme der Brennstoffanalytik. — IfK-Information, Dresden (1983) 120, S. 49—54

Quecksilber in Standardgesteinsproben und die Möglichkeit seiner Attestierung[1])

RUDI SCHEUTZOW, Berlin

Veröffentlichung Nr. 15/88 des Zentralen Geologischen Institutes

Quecksilber dient in der Geochemie als Indikatorelement. Sowohl Quecksilber selbst als auch die Mehrzahl seiner Verbindungen sind starke Gifte. Die Bestimmung von Quecksilber in Gesteinen hat daher für die Geochemie und Umweltfragen große Bedeutung.

Der Clarke-Wert für Quecksilber wird mit 45 bis 67 ng/g angegeben (STOLJAROVA & FILATOVA 1981). Der Hg-Gehalt in Gesteinen liegt also im ng/g-Bereich. Wie in der Spurenelementanalytik überhaupt kommt auch bei der Hg-Bestimmung der Gewährleistung der Richtigkeit durch den Einsatz von Standardproben große Bedeutung zu. Von den im Rahmen der Zusammenarbeit in der Ständigen Kommission für Geologie des RGW bearbeiteten 45 Standardproben ist Quecksilber nur für zwei Erzproben mit höheren Gehalten (Kupfer-Zink-Sulfiderz RUS 1 mit (16 ± 4) ng/g und Kupfer-Kies-Erz

Eingang des Manuskripts in der Redaktion: 2. 11. 1987

[1]) Vortrag, gehalten auf der Laborkonferenz von Intergeotechnika im November 1987 in der VRB

RUS 2 mit (43 ± 4) ng/g attestiert (SCHEUTZOW & KÜHN 1985). Weder der Gehalt noch die Matrix dieser beiden Standardproben entsprechen denen typischer Gesteine.
Quecksilber ist in den Gesteinen gebunden, und zwar vor allem als Sulfid, es kann aber auch elementar enthalten sein. Auf Grund der Flüchtigkeit des Quecksilbers besteht die Gefahr, daß das elementar im Gestein vorliegende Hg entweicht und eine Abnahme der Gehalte verursacht. In diesem Fall wäre eine Attestierung der Hg-Gehalte für Standardproben von Gesteinen unmöglich. Zur Klärung dieser Frage wurden einige Standardproben von Gesteinen auf Beständigkeit der Hg-Gehalte untersucht.
Die Untersuchung erfolgte mit der flammenlosen AAS an einem Eigenbaugerät mit einer Küvette von 20 cm Länge (wie von Dr. Ungethüm im ZGI eingeführt).

Prinzip der Versuchsdurchführung:

Gemessen wird bei 253,7 nm. Die Probe von 3 g wird mit 20 ml 37%iger Salpetersäure und 2 ml Salzsäure (Dichte = 1:1) aufgeschlossen (2 h auf dem Wasserbad) und auf 250 ml im Maßkolben aufgefüllt. Aus dem Maßkolben, in dem sich die ungelöste Substanz abgesetzt hat, wird ein Aliquot für die Bestimmung abgenommen, im Reaktionsgefäß mit $SnCl_2$-Lösung reduziert und das Quecksilber mittels Stickstoffstrom auf einen Goldkollektor überführt. Durch Beheizen des Goldkollektors und den Stickstoffstrom wird das Hg durch die Küvette geleitet und die Absorption gemessen.

Das elementare Quecksilber sollte beim Erwärmen besonders schnell aus dem Gesteinspulver durch Verdampfen entweichen. Daher wurden die Proben 3 h bei 105 °C getrocknet. Unter völlig gleichen Bedingungen wurden getrocknete und ungetrocknete Proben aufgeschlossen und gemessen (Tab. 1). Es fällt auf, daß aus der Standardprobe Kupfererz CuPl beim Erhitzen auf 105 °C kein Hg entweicht. Diese Probe enthält 17% Chalkosin und 1,5% Pyrit. Offenbar liegt das gesamte Hg hier als Sulfid vor. Bei den Standardproben Granit GM, Schwarzschiefer TS und Tonschiefer TB-2 entweichen 20 bis 30% des Quecksilbers. Diese Proben dürfen keinesfalls vor der Hg-Bestimmung getrocknet oder erwärmt werden.
Tab. 2 enthält die Ergebnisse zu den Standardproben Basalt BM, Granit GM, Gneis GnA, Schwarzschiefer TS, Tonschiefer TB-2 und Kupfererz CuPl. Zu diesen Standardproben wurden über mehrere Jahre übereinstimmende Hg-Werte erhalten. In geschlossenen Flaschen sind die Hg-Gehalte also über Jahre stabil, obwohl diese Flaschen im Laufe der Jahre viele Male zur Probenentnahme geöffnet wurden. Bei einigen Proben (TB-2, TS, CuPl) erfolgten die Einwaagen auch aus verschiedenen Flaschen, was keine unterschiedlichen Ergebnisse brachte.
Natürlich sind bei der Untersuchung dieser Standardproben auf Quecksilber für die Attestierung in mehreren Laboratorien verschiedener Länder wegen der kleinen Gehalte und der Problematik der Hg-Bestimmung größere Abweichungen nicht

Tab. 1. Vergleich von Hg-Werten in ng/g aus erhitzten und unbehandelten Proben

	x1 getrocknet (3 h, 105 °C)	x2 nicht getrocknet	$\frac{x1 - x2}{x2} \cdot 100$
GM	15	20	−25
GM	15	21	−29
GnA	5	13	< −62
GnA	5	13	< −62
TS	39	50	−22
TS	38	49	−22
TB2	15	19	−21
TB2	14	18	−22
CuPl	258	258	
CuPl	268	267	

Tab. 2. Hg-Gehalte (x, ng/g) und Standardabweichungen für einige Standardproben

		$\bar{x}$	s	n
BM	1981/1983	26	5	5
GM	1981–1986	21	5	7
GnA	1981–1986	13	2	15[1])
TS	1981–1986	48	7	14[1])
TB-2	1983–1986	15	5	23[1])
CuPl	1976–1987	264	34	12[2])

n – Anzahl der Aufschlüsse

[1]) je Aufschluß 2mal gemessen ($n_i = 2$)

$$s = \sqrt{\frac{2\sum(\bar{x}_i - \bar{x})^2}{n-1}}$$

[2]) Je Aufschluß unterschiedliche Anzahl von Messungen (n_i = 2 bis 10)

$$s = \sqrt{\frac{\sum\limits_{i=1}^{n} n_i(\bar{x}_i - \bar{x})^2}{n-1}}$$

$$\bar{x}_i = \frac{1}{n_i}\sum_{j=1}^{n_i} x_{ij}$$

$$\bar{x} = \frac{1}{N}\sum_{j=1}^{n} n_i\bar{x}_i$$

$$N = \sum_{i=1}^{n} n_i$$

Tab. 3. Orientierende Hg-Werte für einige Standardproben, bestimmt aus einer Einwaage, in ng/g

AN	30	SNS-2	10
AnK	10	SG	6500
BaH	130	ZnU	150
OgG	50	AK	20
GV	15	SW	10
KK	60	FK	20
MV	100	KH	30
MB	20		

auszuschließen. In diesem Fall ist eine Diskussion der Einflußfaktoren in noch stärkerem Maße als bei den übrigen Spurenelementen notwendig, um methodische und apparative Fehlerquellen aufzudecken. Ferner sollten neben einer Wiederholung der Bestimmung im selben Labor gegebenenfalls Kontrolluntersuchungen an Einwaagen aus derselben Flasche in anderen Laboratorien vorgenommen werden. Falls auf diesem Wege große Abweichungen bestätigt werden, enthalten verschiedene Flaschen einer Standardprobe unterschiedliche Hg-Gehalte, und seine Attestierung ist für diese Standardprobe nicht zweckmäßig. Nach den bisherigen Untersuchungen ist dies nicht absolut auszuschließen, jedoch nicht sehr wahrscheinlich.

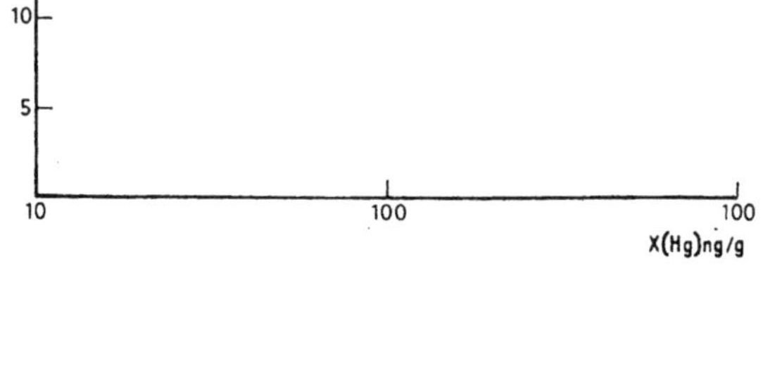

Abbildung. Standardabweichungen in Abhängigkeit vom Hg-Gehalt

Für die Standardproben BM und TS erhielt BEUGE (1976) Werte (20 und 41 ng/g), die denen in Tab. 2 entsprechen. Sein Wert für GM liegt mit 27 ng/g ebenfalls im Bereich des in Tab. 2 angeführten.
Es ist zweckmäßig, die Attestierungsuntersuchungen an den Standardproben BM, TS und CuPl zu beginnen. Diese Proben weisen unterschiedliche Hg-Gehalte auf, die offensichtlich ausreichend stabil sind.
Tab. 3 gibt Auskunft über die Größenordnung der Hg-Gehalte, die in weiteren Standardproben bestimmt wurden. In der Abbildung ist die Standardabweichung der mittels flammenloser AAS erhaltenen Werte ersichtlich, was mit den Literaturdaten übereinstimmt (STOLJAROVA & FILATOVA 1981). Die Reproduzierbarkeit von 30% relativ wird für diesen Konzentrationsbereich von VOLKOVA & OSIKO (1979) als Qualitätsparameter angegeben.

Literatur

BEUGE, P.: Zur Geochemie des Quecksilbers in Magmatiten und Einzelmineralen. – Freiberger Forsch.-H. R. C 313, Leipzig (1976) S. 12
SCHEUTZOW, R.; KÜHN, G.: Die Bearbeitung von Standardproben im ZGI. – Z. angew. Geol., Berlin 31 (1985) 10, S. 305–307
STOLJAROVA, I. A.; FILATOVA, M. P.: Atomno-absorbcionnaja spektrometrija pri analiza mineral'nogo syrja. – Leningrad: Izdat. Nedra 1981
VOLKOVA, G. A.; OSIKO, E. P.: Spravočnye dannye o metodach izmerenija chimičeskogo sostava gornych porod, rud i produktov ich pererabotki. – Moskva: Izdat. Nedra 1979. S. 13

Fehlerbetrachtungen zu markscheiderischen, boden- und gesteinsmechanischen Meßergebnissen und deren Auswirkung auf bergbausicherheitliche Parameter

KARL-HEINZ HEYNE, Leipzig; HORST BOGNITZ, Halle

Zielstellung

In der Anweisung Nr. 6/79 der Obersten Bergbehörde vom 10. 4. 1979 wird u. a. festgelegt, daß „*die vollständige oder teilweise Einbeziehung der Komplexe und Kontrollschwerpunkte in die markscheiderische Sicherheitskontrolle durch die Betriebe unter Berücksichtigung der speziellen bergbausicherheitlichen Bedingungen zu prüfen ist. Die Komplexe und Kontrollschwerpunkte sind durch die Betriebe erforderlichenfalls zu konkretisieren oder zu ergänzen*".
Auf dieser Grundlage werden in Kies-Sand-Tagebauen unter anderem die Komplexe

- Sicherheitspfeiler und Sicherheitsabstände,
- fortschreitende Über- und Unterwasserböschungen,
- bleibende Über- und Unterwasserböschungen,
- Trocken- und Naßkippen,
- Dämme,
- Absperrmaßnahmen,
- Kabel und Leitungen sowie
- Boden- und Gebirgsbewegungen

unter markscheiderische Sicherheitskontrolle genommen.
Grundlage für die Festlegung der Sollwerte dieser Kontrollschwerpunkte sind entweder

- die in Anlage 4 der ABAO 122/1 festgelegten Böschungsparameter oder
- die Ergebnisse von Standsicherheitsnachweisen.

Da als Sollwert eines zu kontrollierenden Komplexes jeweils ein konkreter Wert vorgegeben wird, entstehen für den die Soll- und Istwerte vergleichenden Markscheider dann gewisse Probleme, wenn die Sollwerte im kritischen Bereich nur wenig überschritten werden. Der Markscheider ist in diesem Fall zur notwendigen Belehrung mit entsprechenden Forderungen verpflichtet, obwohl er auf Grund der probabilistischen Sicherheitskonzeption für die Ermittlung der Sollwerte, wie z. B. Böschungs-Neigungswinkel, Sicherheitsabstände usw., nicht von seiner Handlungsweise überzeugt sein kann.
Die vorliegenden Betrachtungen haben, ausgehend von der auf diesem Gebiet bisher erschienenen Fachliteratur, insbesondere von OSWALD, HAGEN und PHILIPP, das Ziel, bei den hierfür zuständigen Aufsichtsorganen und Experten weitere Diskussionen anzuregen.
Hierbei sollte z. B. das Ziel verfolgt werden, für den jeweiligen Sollwert einen Konfidenzbereich anzugeben, der dem Markscheider eine den stochastischen Bedingungen entsprechende Verfahrensweise bei der markscheiderischen Sicherheitskontrolle gestattet.

Bergbau-Sicherheitsparameter beeinflussende Faktoren

Lockergesteinsböschung

Die Stabilität einer Lockergesteinsböschung wird durch folgende Kenngrößen bestimmt:

- geometrische Kenngrößen
  - Böschungswinkel β, in °
  - Böschungshöhe h, in m
  - Grundwasserspiegelhöhe h_w, in m
  - Zugrißtiefe z_0, in m
- bodenmechanische Kenngrößen
  - Reibungswinkel φ', in °
  - Kohäsion c', in kN/m²
  - Rohwichte γ, in t/m³

Festgesteinsböschung

Die Stabilität einer Festgesteinsböschung wird durch folgende Kenngrößen bestimmt:

Bei nur einer parallel zur Böschung streichenden Trennflächenschar

- geometrische Kenngrößen
  - Böschungswinkel β, in °
  - Böschungshöhe h, in m
  - Grundwasserspiegelhöhe h_w, in m
  - Zugrißtiefe z_0, in m
  - Einfallwinkel der Trennfläche ε, in °
- gesteinsmechanische Kenngrößen
  - Reibungswinkel der Trennfläche φ_T, in °
  - Kohäsion der Trennfläche c_T, in kN/m²
  - Rohwichte γ, in t/m³

Bei zwei Trennflächenscharen mit beliebiger Raumstellung zur Böschung

- geometrische Kenngrößen
  - Streichen der Trennfläche A, in °
  - Fallen der Trennfläche A, in °
  - Streichen der Trennfläche B, in °
  - Fallen der Trennfläche B, in °
- gesteinsmechanische Kenngrößen
  - Reibungswinkel der Trennfläche A, in °
  - Reibungswinkel der Trennfläche B, in °

Bei nicht ausgeprägten Trennflächen

- geometrische Kenngrößen
  - Böschungswinkel β, in °
  - Böschungshöhe h, in m
  - Grundwasserspiegelhöhe h_w, in m
  - Zugrißtiefe z_0, in m
- gesteinsmechanische Kenngrößen
  - Reibungswinkel des Gebirgsmassivs φ_M, in °
  - Kohäsion des Gebirgsmassivs c_M, in kN/m²
  - Rohwichte γ, in t/m³

Bei zwei sich schneidenden Trennflächenscharen mit auftretenden Gleitkeilen

- geometrische Kenngrößen
  - Fallrichtung der Trennfläche A, in °
  - Fallwinkel der Trennfläche A, in °

Eingang des Manuskripts in der Redaktion: 23. 4. 1987
[1] Vortrag gehalten auf dem Berg- und Hüttenmännischen Tag Freiberg, Kolloquium 6, am 27. 06. 1986

- Fallrichtung der Trennfläche B, in °
- Fallwinkel der Trennfläche B, in °
- Fallrichtung der Böschung, in °
- Fallwinkel der Böschung, in °
- Böschungshöhe h_A, in m
- Fallrichtung des Terrains, in °
- Fallwinkel des Terrains, in °
- Fallrichtung des Zugrisses, in °
- Fallwinkel des Zugrisses, in °
- Abstand des Zugrisses von der Böschungskante, in m
- Grundwasserspiegelhöhe über Gleitkeilunterkante, in m

— gesteinsmechanische Kenngrößen
- Reibungswinkel der Trennfläche A, in °
- Reibungswinkel der Trennfläche B, in °
- Kohäsion der Trennfläche A, in kN/m²
- Kohäsion der Trennfläche B, in kN/m²
- Rohwichte γ, in t/m³

Streuungen der die Böschungsstabilität bestimmenden Kenngrößen

Stochastischer Charakter der Kenngrößen

Alle eine Böschungsstabilität bestimmenden Kenngrößen sind bei ihrer Bestimmung Zufallswerte und unterliegen den Gesetzen der mathematischen Statistik, wobei zwischen den geometrischen und den gebirgsbeschreibenden Kenngrößen die folgenden grundsätzlichen Unterschiede bestehen:

Eine geometrische Kenngröße ist durch ihren wahren, jedoch unbekannten Wert l definiert. Er wird in der Regel durch eine Anzahl von Messungen l_i ($i = 1 \ldots n$) bestimmt. Der wahrscheinlichste Wert ergibt sich dann durch arithmetische Mittelbildung $\bar{l} = \frac{[l_i]}{n}$ aller Messungen mit einem mittleren Fehler $m = \sqrt{\frac{[vv]}{n-1}}$, wobei $v_i = l_i - \bar{l}$ ist.
Dieser mittlere Fehler charakterisiert den Genauigkeitsgrad bzw. die Fehlerbandbreite *einer* Messung; er entspricht aus mathematischer Sicht sowie der Größe des Wertes der empirischen Standardabweichung bei mathematisch-statistischen Untersuchungen. Die einzelnen Meßergebnisse unterliegen den Gesetzmäßigkeiten der Normalverteilung.
Eine gebirgsbeschreibende Kenngröße wird durch ihren unbekannten Wert der Grundgesamtheit des untersuchten Merkmals x definiert. Er wird mittels einer Stichprobenuntersuchung durch eine Anzahl von Realisierungen x_i($i = 1 \ldots n$) bestimmt und durch seine Momente beschrieben, wobei dem Erwartungswert das arithmetische Mittel $\bar{x} = \frac{[x_i]}{n}$ mit der empirischen Standardabweichung $s = \sqrt{\frac{1}{n-1}\sum_{i=1}^{n}(x_i - \bar{x})^2}$ entspricht.
Die empirische Standardabweichung als stets positiver Wert charakterisiert den Streuungswert einer Stichprobenuntersuchung. Hierbei wird die Gültigkeit einer Normalverteilung angenommen; die geschätzten Parameter werden als Parameter der Normalverteilung angesetzt. Aus diesen Gesetzmäßigkeiten der geometrischen und gebirgsbeschreibenden Kenngrößen werden die folgenden Schlüsse gezogen:

— Geometrische und gebirgsbeschreibende Kenngrößen als die eine Böschungsstabilität bestimmenden Parameter werden als stetige Zufallsgrößen betrachtet und genügen in den meisten Fällen einer Normalverteilung. Sie unterliegen damit den Gesetzen der Stochastik.
— Auf Grund ihrer gleichen mathematischen Definition kann eine komplexe fehlertheoretische Betrachtung der geometrischen und gebirgsbeschreibenden Kenngrößen mit Hilfe des Fehlerfortpflanzungsgesetzes durchgeführt werden. Damit ist es möglich, die Unsicherheiten abgeleiteter bergbausicherheitlicher Parameter abzuschätzen.

Die mittleren Fehler der markscheiderischen Lage- und Höhenmessungen

Auf der Grundlage des im bergmännischen Rißwerk zugelassenen maximalen Lagefehlers $m_L = \pm 1$ m und maximalen Höhenfehlers $m_H = \pm 0{,}2$ m berechnen sich für den mittleren Fehler

— einer Horizontalstrecke $m_{\Delta L} = \sqrt{2} \cdot 1 = \pm 1{,}41$ m
— einer Vertikalstrecke $m_{\Delta H} = \sqrt{2} \cdot 0{,}2 = \pm 0{,}28$ m
— eines Böschungswinkels

$$m_{\tan\beta} = \pm \sqrt{\frac{m_{\Delta H}^2}{\Delta L^2} + \frac{m_{\Delta L}^2 \cdot \Delta H^2}{\cdot \Delta L^4}} = \pm \sqrt{\frac{0{,}08}{\Delta L^2} + \frac{1{,}99 \cdot \Delta H^2}{\Delta L^4}} \quad (1)$$

Unter Anwendung von Formel (1) ergeben sich für bestimmte geometrische Bedingungen die in Tab. 1 zusammengestellten mittleren Böschungswinkelfehler, die z. B. bei einem Böschungswinkel $\beta = 40°$ und den Böschungshöhen H = 20 m, 10 m und 5 m bei $m_\beta = \pm 2{,}9°$, $\pm 5{,}8°$ und $\pm 11{,}4°$ liegen. Bei Echolotmessungen mit dem mittleren Meßfehler von $m_z = \pm 0{,}60$ m ergeben sich für die gleichen geometrischen Bedingungen $m_\beta = \pm 3{,}2°$, $\pm 6{,}4°$ und $\pm 12{,}4°$.

Die Standardabweichungen gesteins- und bodenmechanischer Stichprobenuntersuchungen

Der boden- und gesteinsmechanische Versuch — ein stochastischer Prozeß

Beim boden- und gesteinsmechanischen Versuch ist es nicht möglich, das Ergebnis mit Sicherheit vorherzusagen. Es ist z. B. unsicher, bei welcher Belastungshöhe der Bruch eines Probekörpers eintritt oder welcher Verformungswert einer bestimmten Belastung zuzuordnen ist. Die Ursache für diese „zufälligen Ereignisse" sind die verschiedenartigen Einflußfaktoren, die das Spannungs-Verformungs- und Bruch-Verhalten eines unter Belastung stehenden Gesteinsprobekörpers beeinflussen.
In Tab. 2 sind alle ein boden- oder gesteinsmechanisches Versuchsergebnis im Hinblick auf seine Zufälligkeit mehr oder weniger beeinflussenden Faktoren zusammengestellt. Eine unbedingte Voraussetzung für eine Vergleichbarkeit der im Laboratorium bestimmten gesteinsmechanischen und gesteinsphysikalischen Kennwerte ist die einheitliche Versuchsmethode bezüglich der Versuchs- und Probekörper-Bedingungen. Sind Versuchsergebnisse vergleichbar, so sind ihre Streuungen auf zufällige Versuchs- und Probekörperfehler, insbesondere auf nicht oder nur schwer erfaßbare petrographisch-mineralogische Unterschiede zwischen den zufällig entnommenen Gesteinsprobekörpern zurückzuführen.
Boden- und gesteinsmechanische Kennwerte sind also als Zufallsgröße einer Grundgesamtheit aufzufassen, die durch

Tab. 1. Mittlere Fehler des Böschungswinkels β für bestimmte geometrische Bedingungen und die Genauigkeit des bergmännischen Rißwerkes

Böschungshöhe H, in m	Böschungswinkel β, in °	mittlerer Fehler des Böschungswinkels m_β, in °
10	18,4	±1,1
10	30	±2,9
20	36	±2,8
5	40	±11,4
10	40	±5,8
20	40	±2,9
Für den bei Echoltomessungen anzusetzenden mittleren Messungsfehler $m_L = \pm 0{,}6$ m ergibt sich		
5	40	±12,4
10	40	±6,4
20	40	±3,2

Tab. 2. Einflußfaktoren des Verformungs-, Festigkeits- und Bruchverhaltens von Gesteinen

1. Petrographisch-mineralogische Faktoren

Gefüge (Struktur, Textur, Bindemittel)
Mineralführung (Chemismus)
Porosität und Dichte
Wassergehalt (Feuchtigkeit)
Art und Orientierung von Schwächestellen (Mikro- und Makrorisse)

2. Versuchsbedingte Faktoren

Spannungszustand (ein-, zwei- oder dreiachsig)
Belastungs- (Verformungs-) geschwindigkeit
Temperatur
umgebendes Medium
Belastungs- (Spannungs-) höhe
Belastungsweise (Reihenfolge der Lastaufgabe bei mehrachsigen Versuchen)
Art des Systems „Prüfmaschine — Probekörper" (Federkonstante der Prüfmaschine)

3. Probekörperbedingte Faktoren

Probekörperform
Probekörpergröße
relative Kornfeinheit
Güte der Herstellung des Probekörpers
Probekörperorientierung
Art der Probenahme
latente Spannungen im Material
Größe der Endflächenreibung

ihre Momente ‚Standardabweichung' und ‚Erwartungswert' beschrieben werden. Auf Grund bisheriger Erfahrungen kann eine boden- oder gesteinsmechanische Stichprobe als normalverteilt angesehen werden.

Statistische Auswertung boden- und gesteinsmechanischer Versuchsergebnisse

In den Tab. 3 bis 7 wurde eine Anzahl von Versuchsreihen an verschiedenen Lockerböden und Gesteinen statistisch ausgewertet, wobei eine Trennung nach

— Scher- und Dichteversuchen (Tab. 3)
— einachsigen Druckversuchen (Tab. 4)
— einachsigen Zugversuchen (Tab. 5)
— Biegeversuchen (Tab. 6)
— und dreiachsigen Druckversuchen (Tab. 7)

vorgenommen wurde. Für jede Versuchsreihe mit einer bestimmten Versuchsanzahl n wurden hierbei geschätzt:

— die Standardabweichung s
— der Mittelwert $\bar{x}$
— der Variationskoeffizient $v = \frac{s}{\bar{x}} 100\%$
— das Konfidenzintervall des Mittelwertes (μ), für $\varepsilon = 95\%$
— das Konfidenzintervall der Standardabweichung (σ), für $\varepsilon = 95\%$
— und die maximale zu erwartende Differenz (d) vom wahren Wert bei 5 Versuchen für $\varepsilon = 95\%$

In Tab. 8 sind schließlich die Mittelwerte der vergleichbaren Variationskoeffizienten aus allen Versuchsreihen, getrennt nach den Versuchsarten sowie in Tab. 9 die mittleren Variationskoeffizienten der dreiachsigen Druckversuche zusammengestellt.

Vorherbestimmung der notwendigen boden- und gesteinsmechanischen Versuchsanzahl (Stichprobenanzahl)

Soll der Fehler eines Mittelwertes die vorgeschriebene oder gewählte höchste Abweichung $d = \bar{x} - \mu$ nicht überschreiten, dann läßt sich bei Voraussetzung einer Normalverteilung mit einer statistischen Sicherheit von $\varepsilon = 95\%$ der Verträuensbereich des Mittelwertes

$$\bar{x}_{o,u} = \bar{x} \pm 1{,}96 \frac{s}{\sqrt{n}} \tag{2}$$

ausdrücken durch

$$\bar{x}_{o,u} = \bar{x} \pm d \tag{3}$$

Aus

$$d = 1{,}96 \frac{s}{\sqrt{n}} \tag{4}$$

ergibt sich

$$n = \left(\frac{1{,}96 s}{d}\right)^2 \tag{5}$$

Mit Hilfe der Beziehungen

$$v = \frac{s}{\bar{x}}; \quad s = v \cdot \bar{x} \tag{6}$$

und

$$d' = \frac{d}{\bar{x}}; \quad d = d' \cdot \bar{x} \tag{7}$$

ergibt sich unter Einsetzen von (6) und (7) in (5) die Beziehung

$$n = 3{,}8416 \cdot \frac{v^2}{d'^2}. \tag{8}$$

mittels der man bei bekanntem Variationskoeffizient v und geforderter oder gewünschter maximaler Differenz d vom wahren Wert die notwendige Versuchsanzahl vorausbestimmen kann.
In der Abbildung wird die Beziehung (8) für die statistischen Sicherheiten $\varepsilon_1 = 95\%$ und $\varepsilon_2 = 99\%$ dargestellt.
In (2) bis (8) bedeuten:

x — Mittelwert
s — Standardabweichung
n — Stichprobenanzahl
v — Variationskoeffizient
d' — gewählte maximale prozentuale Differenz vom Mittelwert.

Schlußfolgerungen aus den Ergebnissen

Unter Ansatz der ermittelten und in den Tab. 8 und 9 zusammengestellten Variationskoeffizienten sowie einer maximalen Differenz zwischen Versuchsergebnissen und wahrem Wert von 10% ergeben sich für $\varepsilon = 95\%$ folgende Versuchsanzahlen:

— einachsiger Druckversuch Salzgestein:	n = 15 Versuche
— einachsiger Druckversuch Nichtsalzgestein:	n = 5 Versuche
— einachsiger Zugversuch Salzgestein:	n = 34 Versuche
— einachsiger Zugversuch Nichtsalzgestein:	n = 3 Versuche
— Biegeversuch Nichtsalzgestein:	n = 2 Versuche
— dreiachsiger Druckversuch Salzgestein:	n = 2 Versuche
— dreiachsiger Druckversuch Nichtsalzgestein:	n = 1 Versuch
— Scherversuch zur Bestimmung von	
• φ'	n = 10 Versuche
• c'	n = 47 Versuche
— Dichtebestimmung γ	n = 1 Versuch

Diese Werte sind als Empfehlungen für die Planung boden- und gesteinsmechanischer Versuche anzusehen. Tatsächlich sollten jedoch nach Durchführung einer jeden Versuchsreihe die statistischen Maßzahlen erneut bestimmt und daraus die endgültige Versuchsanzahl nach Gl. (8) berechnet werden, wobei als statistische Sicherheit $\varepsilon = 95\%$ und als zu wählende Maximaldifferenz d = 10% angesetzt werden sollten.
Zu beachten ist, daß der Variationskoeffizient bei dreiachsigen gesteinsmechanischen Druckversuchen stets bedeutend niedriger liegt als bei einachsigen Druckversuchen.
Bei den bodenmechanischen Versuchen ist stets der Kennwert c' mit dem größten Fehler behaftet (Tab. 3).

Ergebnisse von fehlertheoretischen Betrachtungen zur Auswirkung von Fehlern geometrischer und gebirgsbeschreibender Kennwerte auf die Unsicherheit bergbausicherheitlicher Parameter

Böschungssicherheit einer Kies-Sand-Böschung

Gemäß der Sicherheitsbeziehung

$$S = \frac{\tan \varphi'}{\tan \beta} \tag{9}$$

Tab. 3. Streuung der Ergebnisse von Scherversuchen und Dichtebestimmungen an verschiedenen Lockerböden

Lockerb.den, Fundort	n	s	$\bar{x}$	v in %	< μ <	< σ <	d bei n = 5	in %
Geschiebemergel	10	4,644	29,700	15,6	26,8–32,6	3,2–8,5	4,1	13,7
Pomßen; Φ, in °	5	5,541	29,200	19,0	24,3–34,1	3,3–15,9	4,9	16,6
	5	4,147	30,200	13,7	26,6–33,8	2,5–11,9	3,6	12,0
Geschiebemergel	10	0,667	1,897	35,2	1,5–2,3	0,5–1,2	0,6	30,8
Nobitz; c, in kN/m²	5	0,562	1,704	33,0	1,2–2,2	0,3–1,6	0,5	28,9
	5	0,770	2,090	36,8	1,4–2,8	0,5–2,2	0,7	32,3
Geschiebemergel	10	0,086	2,030	4,2	2,0–2,1	0,0–0,2	0,1	3,7
Nobitz; γ, in t/m³	5	0,068	2,034	3,3	2,0–2,1	0,0–0,2	0,1	2,9
	5	0,110	2,026	5,4	1,9–2,1	0,1–0,3	0,1	4,8
Bänderton Leipzig; φ_g, in °	7	1,781	8,257	21,6				
Bänderton Leipzig; c_g, in kN/m²	7	2,640	4,114	64,2				
Bänderton Leipzig; $S_{Gerät}$	7	0,050	1,175	4,3				
Bänderton Leipzig; $S_{Böschung}$	7	0,066	1,142	5,8				

Die Kennzahlen des Leipziger Bändertones wurden von sieben verschiedenen Labors ermittelt!

Tab. 4. Streuung der Ergebnisse von einachsigen Druckversuchen an verschiedenen Gesteinen

Gestein/Fundort		n	s in N/mm²	$\bar{x}$ in N/mm²	v in %	< μ < in N/mm²	< σ < in N/mm²	d bei n = 5 in N/mm²	in %
Hartsalz/Bischofferode	σ_{DZ}	15	6,837	34,323	19,9	30,6–38,1	5,0–10,8	6,0	17,5
Hartsalz/Bischofferode	σ_{DZ}	50	6,103	40,564	15,0	38,9–42,3	5,0–7,6	5,4	13,2
Hartsalz/Bischofferode	σ_{DZ}	36	10,591	39,423	26,9	35,8–43,0	8,4–13,8	9,3	23,6
Steinsalz/Bernburg	σ_{DZ}	38	5,984	36,173	16,5	34,2–38,1	4,9–7,7	5,3	14,6
Granit/Blackingstone	σ_{DZ}	10	32,176	179,580	17,9	156,9 bis 202,3	23,5–58,8	28,3	15,8
Granit/Westerly	σ_{DZ}	7	10,340	256,331	4,0	246,0 bis 266,7	6,7–22,7	9,1	3,6
Quarzit/Witwatersrand	σ_{DZ}	5	15,253	212,948	7,2	194,0 bis 231,9	9,2–43,9	13,4	6,3
Marmor/?	σ_{DW}	4	6,547	58,104	11,3	47,7–68,5	3,7–24,4	5,8	9,9
Buntsandstein/?	σ_{DW}	4	3,333	27,201	12,3	21,9–32,5	1,9–12,4	2,9	10,8
Kalkstein/Solnhofen	σ_{DW}	3	12,648	224,900	5,6	193,5–256,3	6,6–79,2	11,1	5,0
Basalt/?	σ_{DW}	3	33,852	310,054	10,9	225,9 bis 394,2	9,2–212,0	29,8	9,6
Kalkstein/Portland	σ_{DZ}	2	17,607	54,750	32,2	–213,3	7,9–562,4	15,5	28,3
Sandstein/Dershire	σ_{DZ}	13	8,050	52,654	15,3	47,8–57,5	5,8–13,3	7,1	13,5
Dolerit/Northumberland	σ_{DZ}	15	35,968	286,600	12,5	266,7 bis 306,5	26,3–56,7	31,7	11,0
Sandstein/Pennant Wales	σ_{DZ}	3	1,069	196,967	0,5	194,3 bis 199,6	0,6–6,7	0,9	0,5
Sandstein/Darley Dale	σ_{DZ}	5	3,286	80,080	4,1	75,0–84,2	2,0–9,4	2,9	3,6
Marmor/Carrara	σ_{DZ}	3	1,852	92,400	2,0	87,8–97,0	1,0–11,6	1,6	1,8

σ_{DZ} – einachsiger Zylinder-Druckversuch σ_{DW} – einachsiger Würfel-Druckversuch

Tab. 5. Streuung der Ergebnisse von einachsigen Zugversuchen an verschiedenen Gesteinen

Gestein/Fundort		n	s in N/mm²	$\bar{x}$ in N/mm²	v in %	< μ < in N/mm²	< σ < in N/mm²	d bei n = 5 in N/mm²	in %
Hartsalz/Bischofferode	σ_{ZS}	10	0,620	2,452	25,3	2,0–2,9	0,4–1,1	0,6	22,3
Steinsalz/Bernburg	σ_Z	15	0,436	1,302	33,5	1,1–1,5	0,3–0,6	0,4	29,5
Marmor/?	σ_Z	5	1,022	6,433	15,9	5,2–7,7	0,6–2,9	0,9	14,0
Buntsandstein/?	σ_Z	4	0,131	1,428	9,2	1,2–1,6	0,1–0,5	0,1	8,1
Kalkstein/Solnhofen	σ_Z	3	0,749	10,232	7,3	8,4–12,1	0,4–0,7	0,7	6,4
Basalt/?	σ_Z	2	0,485	19,270	2,5	14,9–23,6	0,2–15,5	0,4	2,2

σ_Z – einachsiger Zugversuch σ_{ZS} – Scheibendruckversuch

Tab. 6. Streuung der Ergebnisse von Balken-Biegeversuchen an verschiedenen Gesteinen

Gestein/Fundort		n	ŝ in N/mm²	$\bar{x}$ in N/mm²	v in %	< μ < in N/mm²	< σ < in N/mm²	d bei n = 5 in N/mm²	in %
Marmor/?	σ_B	3	0,538	7,829	6,9	6,5–9,2	0,3–3,4	0,5	6,1
Buntsandstein/?	σ_B	4	0,039	2,420	1,6	2,4–2,5	0,0–0,2	0,03	1,3
Kalkstein/Solnhofen	σ_B	4	2,750	17,676	15,6	13,3–22,0	1,6–10,3	2,4	13,7
Basalt/?	σ_B	4	0,747	24,320	3,1	23,1–25,5	0,4–2,8	0,7	2,7

σ_B – Balken-Biegeversuch

Tab. 7. Streuung der Ergebnisse von dreiachsigen Druckversuchen an verschiedenen Gesteinen

Gestein/Fundort		n	s in N/mm²	$\bar{x}$ in N/mm²	v in %	< μ < in N/mm²	< σ < in N/mm²	d bei n = 5 in N/mm²	in %
Quarzit/Witwatersrand	σ_{DZ}	5	15,253	212,948	7,2	194,0–231,9	9,2–43,9	13,4	6,3
„	$\sigma_{T/8,3}$	5	7,150	284,115	2,5	270,4–297,8	4,3–20,6	6,3	2,2
„	$\sigma_{T/11,2}$	3	8,052	358,362	2,2	338,4–378,4	4,2–50,4	7,1	2,0
Hartsalz/Bischofferode	σ_{DZ}	15	6,837	34,323	19,9	30,6–38,1	5,0–10,8	6,0	17,5
„	$\sigma_{T/2,5}$	3	5,433	56,290	9,7	42,8–69,8	2,8–34,0	4,8	8,5
„	$\sigma_{T/4,0}$	3	3,736	66,783	5,6	57,5–76,1	1,9–23,4	3,3	4,9
„	$\sigma_{T/9,8}$	3	3,226	83,945	3,8	75,9–92,0	1,7–20,2	2,8	3,4
„	$\sigma_{T/14,1}$	3	6,114	95,223	6,4	80,0–110,4	3,2–38,3	5,4	5,7
„	$\sigma_{T/19,6}$	3	9,852	101,401	9,7	76,9–125,9	5,1–61,7	8,7	8,6
„	$\sigma_{T/24,5}$	3	2,887	114,640	2,5	107,5–121,8	1,5–18,1	2,5	2,2
„	$\sigma_{T/29,4}$	3	7,474	118,170	6,3	99,6–136,7	3,9–46,8	6,6	5,6
„	$\sigma_{T/34,5}$	3	7,134	126,310	5,6	108,6–144,0	3,7–44,7	6,3	5,0

σ_{DZ} – einachsiger Zylinder-Druckversuch
$\sigma_{T/8,3}$ – dreiachsiger Druckversuch bei 8,3 N/mm² Mantelspannung.

Tab. 8. Mittlere Variationskoeffizienten von verschiedenen boden- und gesteinsmechanischen Versuchsarten

Versuchsart	Mittlerer Variationskoeffizient		
	für alle Gesteine oder Böden in %	für Salzgesteine in %	für Nichtsalzgesteine in %
einachsiger Druck σ_D	10,3	19,6	10,5
einachsiger Zug σ_Z	15,6	29,4	8,7
Biegung σ_B	6,8	–	6,8
dreiachsiger Druck σ_T	5,4	6,2	2,3
Scherversuch			
φ'	16,1	–	–
c'	35,0	–	–
Dichtebestimmung γ	4,3	–	–

Tab. 9. Vergleich der Variationskoeffizienten von ein- und dreiachsigen Druckversuchen

Gestein	einachsiger Druckversuch v_1 in %	dreiachsiger Druckversuch v_3 in %	$\frac{v_3}{v_1}$
Quarzit Witwatersrand	7,2	2,3	0,32
Hartsalz Bischofferode	19,6	6,2	0,32

und der daraus abgeleiteten Fehlergleichung

$$m_s = \sqrt{\frac{m^2_{\tan\varphi'}}{\cos^4\varphi' \cdot \tan^2\beta} + \frac{m^2_{\tan\beta'} \cdot \tan^2\varphi'}{\tan^4\beta \cdot \cos^4\beta}} \tag{10}$$

ergibt sich für die Kennwerte

$\varphi' = 38°$ $\quad s\varphi' = 4{,}78°$
$\beta = 30°$ $\quad m\beta = \pm 5{,}8°$

eine Böschungs-Standsicherheit $S = 1{,}35 \pm 0{,}39$; d. h., der Variationskoeffizient beträgt in diesem Fall $V = 28{,}9\%$, und im ungünstigen Falle wird die Grenzfestigkeit unterschritten.
Für die Kennwerte

$\varphi' = 42°$ $\quad s\varphi' = 4{,}78°$
$\beta = 36°$ $\quad m\beta = \pm 3{,}00°$

ergibt sich eine Böschungs-Standsicherheit $S = 1{,}24 \pm 0{,}25$; auch hier wird im ungünstigen Fall die Grenzfestigkeit bei dem Variationskoeffizienten $V = 20{,}2\%$ unterschritten.

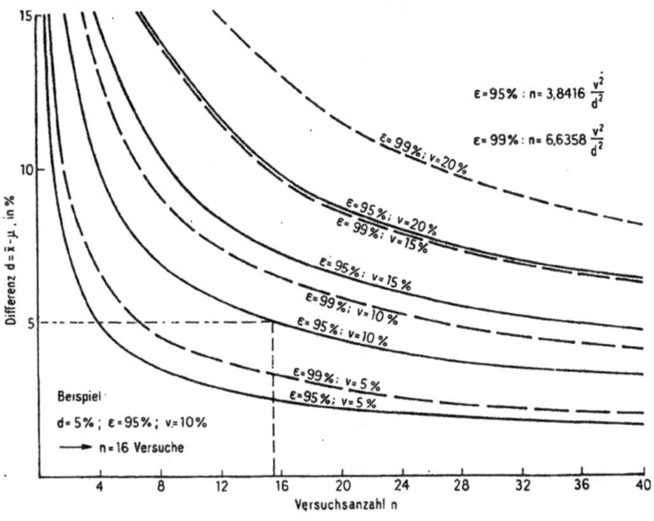

Abb. Abhängigkeit zwischen dem Variationskoeffizienten v, dem halben Konfidenzbereich d und der Stichprobenanzahl n für die statistischen Sicherheiten $\varepsilon_1 = 95\%$ und $\varepsilon_2 = 99\%$

Näherungsverfahren nach HOEK

Lockergesteinsböschung

Gemäß den Funktionen

$$X = \beta - 1{,}2 \cdot \varphi' \tag{11}$$

$$Y = \frac{\gamma \cdot h}{c'} \tag{12}$$

und den daraus abgeleiteten Fehlergleichungen

$$m_x = \sqrt{m\beta^2 + 1{,}44 \cdot m_{\varphi'}^2} \tag{13}$$

$$m_y = \frac{1}{c}\sqrt{h^2 \cdot m_\gamma^2 + \gamma^2 \cdot m_h^2 + \frac{\gamma^2 \cdot h^2}{c^2} \cdot m_c^2} \tag{14}$$

ergeben sich für die Kennwerte

$\varphi' = 30°$ $\quad s_{\varphi'} = 4{,}64°$
$\beta = 45°$ $\quad m_\beta = \pm 2{,}0°$
$\gamma = 1{,}8\ t/m^3$ $\quad s_\gamma = 0{,}086\ t/m^3$
$c = 1{,}9\ kN/m^2$ $\quad s_c = 0{,}6665\ kN/m^2$
$h = 10\ m$ $\quad m_h = \pm 0{,}2828\ m$

die Funktionswerte

$X = 9{,}0° \pm 5{,}9°$ $\quad (V = 65{,}6\%)$
$Y = 9{,}5° \pm 3{,}4°$ $\quad (V = 35{,}5\%)$

und davon abgeleitet die Böschungs-Standsicherheit $S = 1{,}60 \pm 0{,}45$; der Variationskoeffizient liegt also bei $V = 28{,}1\%$.

Festgesteinsböschung

Für die Kennwerte

$h = 7{,}5\ m$ $\quad m_h = \pm 0{,}2828\ m$
$h_w = 5{,}5\ m$ $\quad m_{h_w} = \pm 0{,}2828\ m$
$\gamma = 2{,}04\ t/m^3$ $\quad s_\gamma = 0{,}1\ t/m^3$
$\varphi' = 22°$ $\quad s_{\varphi'} = 5°$
$c' = 0{,}5\ kN/m^2$ $\quad s_{c'} = 0{,}15\ kN/m^2$
$S = 1{,}1$

ergeben sich mittels der Funktionen

$$\beta = X + \varphi'\left(1{,}2 - 0{,}3\,\frac{h_w}{h}\right) \tag{15}$$

und (12) und den daraus abgeleiteten Fehlergleichungen

$$m_\beta = \sqrt{m_x^2 + \left(1{,}2 - \frac{0{,}3 \cdot h_w}{h}\right)^2 \cdot m_\varphi^2 + \left(\frac{0{,}3 \cdot \varphi}{h}\right)^2 \cdot m^2_{h_w} + \left(\frac{0{,}3 \cdot \varphi \cdot h_w}{h^2}\right)^2 m_h^2} \tag{16}$$

und (14)

$Y = 30{,}6$ $\quad S_y = 9{,}4$ $\quad (V = 30{,}7\%)$
Aus $Y = 30{,}6$ und $S = 1{,}1$: $\quad X = 5{,}4$ $\quad S_x = 3{,}57$ $\quad (V = 65{,}6\%)$
$\beta = 27{,}0° \pm 3{,}6°$ $\quad (V = 13{,}3\%)$.

Im Ergebnis der Untersuchung ist festzustellen, daß die mit einer Sicherheit von $S = 1{,}1$ berechnete Böschung des Winkels $\beta = 27{,}0°$ mit einem mittleren Fehler $m_\beta = \pm 3{,}6° \mathrel{\hat=} 13{,}3\%$ behaftet ist.

Lamellen-Verfahren nach KREY

Beim Lamellen-Verfahren nach KREY berechnet sich die Böschungsstandsicherheit einer Lamelle mit der Beziehung

$$\eta_s = \frac{1}{G \cdot \sin\alpha}\left(\frac{b \cdot c'}{\cos\alpha} + G \cdot \cos\alpha \cdot \tan\varphi'\right). \tag{17}$$

Setzt man für $G = h \cdot b \cdot \gamma$ und $\sin\alpha = \frac{1}{10}; \frac{5}{10}$ und $\frac{9}{10}$ ein, so erhält man die Beziehungen

$$\eta_{s_1} = 10{,}05 \frac{c'}{h \cdot \gamma} + 9{,}95 \cdot \tan \varphi' \quad (18)$$

$$\eta_{s_2} = 2{,}31 \frac{c'}{h \cdot \gamma} + 1{,}73 \cdot \tan \varphi' \quad (19)$$

$$\eta_{s_3} = 2{,}55 \frac{c'}{h \cdot \gamma} + 0{,}48 \cdot \tan \varphi' \quad (20)$$

mit den Fehlergleichungen

$$m_{\eta_{st}} = \sqrt{A^2 \frac{m_{c'}^2}{h^2 \cdot \gamma^2} + A^2 \frac{c'^2 \cdot m_h^2}{h^4 \cdot \gamma^2} + A^2 \frac{c'^2 \cdot m_\gamma^2}{h^2 \cdot \gamma^4} + B^2 \frac{m_{\varphi'}^2}{\cos^4 \cdot \varphi'}} \quad (21)$$

A = 10,05; 2,31; 2,55
B = 9,95; 1,73; 0,48.

Im Ergebnis der Berechnungen ergeben sich folgende Sicherheiten:

$\eta_{s_1} = 6{,}70 \pm 1{,}13$ ($\triangleq$ 16,8%)
$\eta_{s_2} = 1{,}22 \pm 0{,}20$ ($\triangleq$ 16,6%)
$\eta_{s_3} = 0{,}52 \pm 0{,}10$ ($\triangleq$ 19,3%)

Hierbei werden folgende Kennwerte verwendet:

$h = 10$ m	$m_h = \pm 0{,}2828$ m
$\gamma = 2{,}0$ t/m³	$s_h = 0{,}086$ t/m³
$c' = 1{,}9$ kN/m²	$s_{c'} = 0{,}6665$ kN/m²
$\varphi' = 30°$	$s_{\varphi'} = 4{,}64°$

Der Variationskoeffizient liegt in diesem Fall zwischen 17% und 20%, im Mittel bei 17,6%.

Zusammenfassung der Ergebnisse

Bei der Durchführung boden- und gesteinsmechanischer Versuche sind folgende Erkenntnisse zu beachten:

- Boden- und gesteinsmechanische Versuche unterliegen den Gesetzen der Stochastik und sind entsprechend zu planen, durchzuführen und auszuwerten.
- Stichproben boden- und gesteinsmechanischer Ermittlungen sind normal verteilt.
- Wichtigstes Gütemaß einer boden- oder gesteinsmechanischen Stichprobe ist der Variationskoeffizient.
- Mit Hilfe des Variationskoeffizienten v und der zu wählenden maximalen Differenz zwischen Versuchswert und wahrem Wert ist die Stichprobenanzahl für die statistische Sicherheit von 95% nach Gl. (8) zu errechnen.
- Der mittlere Variationskoeffizient bodenmechanischer Versuche wurde für
  - den Reibungswinkel φ' mit v = 16,1%
  - die Kohäsion c' mit v = 35,0%
  - die Rohdichte γ mit v = 4,3%

  bestimmt.
- Der mittlere Variationskoeffizient gesteinsmechanischer einachsiger Druckversuche wurde für
  - Nichtsalzgesteine mit v = 10,5%
  - Salzgesteine mit v = 19,6%

  bestimmt.
- Der mittlere Variationskoeffizient gesteinsmechanischer dreiachsiger Druckversuche wurde für
  - Nichtsalzgesteine mit v = 2,3%
  - Salzgesteine mit v = 6,2%

  bestimmt.
- Der Variationskoeffizient bei dreiachsigen gesteinsmechanischen Druckversuchen liegt erheblich niedriger als bei einachsigen gesteinsmechanischen Versuchen und wurde sowohl für Salzgestein als auch für Nichtsalzgestein zu 32% der Variationskoeffizienten der einachsigen Versuche bestimmt.
- Bei den bodenmechanischen Versuchen ist die Kohäsion c' mit den größten Streuungen behaftet (v = 35,0%); bei den gesteinsmechanischen einachsigen Zugversuchen an Salzgesteinen ist mit einer hohen Fehlerquote (v = 29,4%) zu rechnen, während die Zug- und Biegeversuche an Nichtsalzgesteinen geringere Streuungswerte (v = 8,7%; v = 6,8%) aufweisen.
- Aufbauend auf den Genauigkeitsvorgaben für das bergmännische Rißwerk beträgt der mittlere Fehler einer horizontalen Land- oder Wasserstrecke $m_L = \pm 1{,}41$ m, der mittlere Fehler einer Vertikalstrecke auf dem Land $m_H = \pm 0{,}28$ m und im Wasser $m_H = \pm 0{,}63$ m. Daraus leiten sich fehlertheoretische mittlere Fehler des Böschungswinkels in der Größenordnung von $m_\beta = \pm 3°$ bis $\pm 12°$ ab.
- Mittels fehlertheoretischer Untersuchungen wurden auf der Grundlage der nachgewiesenen Streuungs- bzw. Fehlerwerte geometrischer und gebirgsbeschreibender Kennwerte die Fehlerquoten bergbausicherheitlicher Parameter berechnet.
  In Abhängigkeit von den geometrischen Bedingungen wurden Variationskoeffizienten der Sicherheitsparameter von v = 13% bis v = 29%, im Mittel von v = 22% berechnet.

Zusammenfassung

Auf der Grundlage von Erkenntnissen über Streuungswerte der Gaußschen Fehlertheorie sowie der mathematischen Statistik ist eine Abschätzung der zu erwartenden Unsicherheiten von markscheiderischen, boden- und gesteinsmechanischen Meß- und Berechnungsergebnissen möglich. Daraus werden Schlußfolgerungen sowohl für die Wertung solcher Meß- bzw. Versuchsergebnisse als auch für die Bewertung geotechnischer Berechnungsergebnisse für die Bergbausicherheit (markscheiderische Sicherheitskontrolle) gezogen.

Резюме

На основе данных о величине разброса значений по гауссовой теории ошибок, а также математической статистики возможна оценка ожидаемых погрешностей результатов маркшейдерских измерений и данных по механическим свойствам грунтов и пород. На этой основе делаются выводы относительно этих результатов, а также выводы для оценки результатов геотехнических расчетов по безопасности горных работ (маркшейдерский контроль безопасности).

Summary

Based on Gaussian error theory as well as scatter value findings and mathematical statistics it is possible to estimate the expected inaccuracy of surveying, soil-mechanical and petrophysical measuring and calculation results. From this, conclusions can be drawn for the evaluation of surveying, soil-mechanical and petrophysical measuring and tests results and for the assessment of geotechnical calculation results concerning safety in mines.

Literatur

FÖRSTER, W.; HAGEN, C.: Zur Genauigkeit bodenmechanischer Aussagen bei der Untersuchung von Böschungen und zur Verantwortung des Sachverständigen. – Z. angew. Geol., Berlin **25** (1978) 12, S. 624–631

HAGEN, C.: Beitrag zum Aufbau einer probabilistischen Sicherheitskonzeption für die Dimensionierung von Böschungen im Lockergestein. – Freiberger Forsch.-H. R. A **708**, Leipzig (1985)

– Beitrag zum Aufbau einer probabilistischen Sicherheitskonzeption für die Dimensionierung von Böschungen im Lockergestein. – Neue Bergbautechnik, Leipzig **14** (1984) 8, S. 287–292

HANISCH, J.; STRUCK, W.: Charakteristischer Wert einer Boden- oder Materialeigenschaft aus Stichprobenergebnissen und zusätzlicher Information. – Bautechnik, Berlin(West) **62** (1985) 10, S. 338–348

JUNGHANS, H.; MANHENKE, V.; PESTER, L.: Untersuchungen zur notwendigen Genauigkeit des geometrischen Lagerstättenmodells von Braunkohlenfeldern aus bodenmechanischer Sicht. – Neue Bergbautechnik, Leipzig **14** (1984) 3, S. 81–84

MANHENKE, V.; PESTER, L.: Die Zuverlässigkeit von Bohrergebnissen zur Bestimmung der Geometrie von Braunkohlenlagerstätten. – Neue Bergbautechnik, Leipzig **15** (1985) 1, S. 21–27

MARTENS, H.-J.: Meßfehler und Meßunsicherheit. – Vermessungstechnik, Berlin **33** (1985) 1, S. 17–20

MOLEK, H.; REUTER, F.: Probleme bei der Erfassung repräsentativer Kluftparameter als Grundlage für Standsicherheitsbeurteilungen von Felsböschungen. – Z. angew. Geol., Berlin **25** (1979) 10, S. 494–497

MÜLLER, B.; KLENGEL, K.-J.: Praktische Ergebnisse bei der Aufnahme und Bestimmung ingenieurgeologisch-felsmechanischer Kennwerte zur Bewertung der Standfestigkeit von Fels. – Z. angew. Geol., Berlin **25** (1979) 10, S. 484–493

OSWALD, O.-G.: Beitrag zum Problem des Sicherheitsfaktors bei der Bemessung von Böschungen im Lockergestein. – Freiberger Forsch.-H., R. A **595**, Leipzig (1979)

PHILIPP, H.: Gegenwärtiger Stand der Scherfestigkeitsermittlung in den Laboratorien des Industriezweiges Braunkohle. — Vortrag, IV. Bodenmechanisches Kolloquium, 21. u. 22. 11. 1984 in Cottbus
REISNER, H.; PARELKA, F.: Die Anwendung wissenschaftlich-technischer Ergebnisse zur Gewährleistung der geotechnischen Sicherheit in den Tagebauen. — Neue Bergbautechnik, Leipzig 15 (1985) 4, S. 137—138
REISSMAN, G.: Was ist der Unterschied zwischen dem mittleren Fehler und der Standardabweichung? — Vermessungstechnik, Berlin 31 (1983) 11, S. 382—383
Anweisung Nr. 6/79 des Leiters der Obersten Bergbehörde der DDR vom 4. 10. 1979
Arbeits- und Brandschutz-Anordnung 1221/1 — Sicherheit im Bergbau über Tage vom 5. 10. 1973. GBl.-Sonderdr. Nr. 768 vom 15. 3. 1974

Tagungsberichte

INQUA 87 Canada

XIIth International Congress/Programme with Abstracts

Ottawa: International Union for Quaternary Research, National Research Council of Canada 1987. 311 S., $ 40.—

Der Band enthält das zum Kongreß in Kanada überreichte Programm und 1085 Kurzfassungen der Vorträge. Von den Symposien (7) und Vortragsveranstaltungen (49) seien auszugsweise genannt:

Symposien:

A: Beginn, Wachstum und Zerfall des Laurentiden-Eisschildes
B: Glaziomarine Umwelt (Faziesräume)
E: Angewandte Quartärforschung

Allgemeine Vortragssitzungen:

5: Grundwasser
13: Glaziale und glazifluviatile Prozesse

Spezielle Vortragsveranstaltungen:

5: Globaler Kohlenstoffzyklus — paläoklimatische Perspektiven
8: Mineral- (Rohstoff-) Erkundung in (ehemals) vereisten Gebieten
16: Meeresspiegel: Geodynamik und Klima
18: Globale Veränderungen
24: Paläoerdbebenhäufigkeit und Neotektonik

Es ist natürlich nicht möglich, an dieser Stelle auf einzelne Themen einzugehen. Einen Einblick in die Vortragskurzfassungen zeigt aber deren aktuelle Gruppierung:

- Oberflächenkartierung mit Hilfe von Satelliten- und Luftbildern incl. solcher Versuche wie Vergleich der borealen Waldausdehnung und Baumringmessungen zur Beurteilung der CO_2-Erhöhung in der Atmosphäre;
- geologische, paläontologische, geochemische und geophysikalische Arbeiten, um auf der Basis zeitlich möglichst detaillierter paläoklimatischer Analysen Vorhersagen für die klimatische Entwicklung der näheren Zukunft (XXI. Jahrhundert) treffen zu können (Temperaturen, Niederschläge, Meeresspiegelschwankungen, Überschwemmungen in Küstenniederungen, Bodenerosion);
- geochemische und schwermineralogische Untersuchung von Grundmoränen (Tills) und anderen glazialen Ablagerungen incl. der Rekonstruktion von Transportrichtungen zur Prospektion auf Erze, Edelmetalle und Diamanten;
- Verfeinerung der Grundmoränenstratigraphie hinsichtlich hydrogeologischer Prognosen über das Auftreten und die Lagerungsverhältnisse von Intertill-Aquiferen;
- Klärung quartärer Schichtenfolgen und ihrer paläohydrologischen Daten in Afrika im Hinblick auf bessere Grundwasserversorgungen;
- detaillierte Quartärkartierungen und spezielle Untersuchungen zur Rutschungsgefährdung in Kohletagebauen Kanadas sowie von Erdrutschgefährdungen in besiedelten Gebieten,
- genetische Till-Klassifikationen und Beziehungen zu geotechnischen Eigenschaften der Tills im Hinblick auf ingenieurgeologische Aufgaben,
- Permafrostprobleme im Hinblick auf Baugrund und Agrikultur incl. Pipeline-Bau in derartigen Gebieten.

Von Bedeutung sind auch die Vorstellungen neuer Übersichtskarten wie z. B. der „Kryolithischen Karte der UdSSR" i. M. 1:2500000 sowie der „Quartärkarte von China und angrenzender Meeresgebiete" im gleichen Maßstab.

Aus der Fülle der weiteren Untersuchungsergebnisse können nur bestimmte Tendenzen herausgestellt werden, wie die weitere Zunahme der Forschungstätigkeit im Quartär der südlichen Hemisphäre, die Weiterführung der Quartärbearbeitung von Bohrungen in verschiedenen Meeresgebieten aber auch die Zunahme von Bohrungen in tiefen Seen bzw. Seen mit Hunderten von Metern mächtigen Quartärablagerungen, um eine möglichst vollständige Aufzeichnung der starken Klimaänderungen im Quartär zu erfassen, das Ringen um immer genauere Methoden der physikalischen Altersdatierung quartärer Bildungen, um zu exakten Korrelierungen und damit zum besseren Verständnis der geologischen Vorgänge in der Zeit zu gelangen (neu: Optische Datierung sowie Einzelkorn-$Ar^{40/39}$-LASER-Datierung), die Quantifizierung der Merkmale quartärer Ablagerungen durch geochemische, paläontologische und andere Daten und damit ihre besseren Vergleichsmöglichkeiten hinsichtlich genetischer, faziell-ökologischer, paläoklimatischer und anderer Aussagen sowie das ständige Bemühen um eine detailliertere stratigraphische Untergliederung und Korrelation.

Den erreichten Fortschritt auf dem Gesamtgebiet der Quartärforschung — soweit er sich in den angemeldeten Vorträgen widerspiegelt — den im Quartär arbeitenden Fachkollegen zugänglich gemacht zu haben, verdient Anerkennung für die Herausgeber.

A. G. CEPEK

...

Recognition of Uranium Provinces

Vienna: International Atomic Energy Agency 1988. 462 S., 17 Beitr., zahlr. Abb., Tab., Lit., ÖS 940.—

(Panel Proceedings Serie)

Das Buch enthält die Protokolle der von der Internationalen Atomenergiebehörde organisierten und vom 18. bis 20. September 1985 in London abgehaltenen internationalen Tagung zum Kenntnisstand der Uranprovinzen, einschließlich einiger Auszüge aus der anschließenden Diskussion. Den 50 Teilnehmern dieser Beratung aus 23 größtenteils OECD-Ländern wurden Beiträge zu drei Themenkreisen angeboten: zur Entstehung von Uranprovinzen, zu ihren Charakteristika und zu ihrer Erkundung.

Die vier Beiträge zum ersten Themenkomplex beschäftigen sich mit der globalen Uranmigration von der Entstehung des Planeten Erde bis zur Gegenwart. Die bedeutendsten Uranprovinzen und Lagerstättentypen der Welt werden aufgeführt.

Den Hauptteil des Buches nehmen die zwölf Beiträge über die wichtigsten Uranlagerstätten und ihre Charakteristika ein. Beschrieben werden fast ausschließlich Lagerstätten in präkambrischen Gesteinskomplexen. Einzige Ausnahme sind die Vorkommen vom Colorado-Plateau. Das Spektrum der Lagerstätten reicht von den bekannten Vorkommen in Kanada, im südlichen Afrika und in Australien bis zu den weniger bekannten Uranprovinzen Brasiliens, Indiens und Grönlands.

Ein Beitrag zum dritten Themenkomplex befaßt sich in verallgemeinerter Form mit Problemen der Suche und Erkundung in verschiedenen Uranprovinzen der Welt.

Die abschließend dargelegte Diskussion beginnt mit einem allgemeinen Teil zu einigen Details bezüglich bestimmter Lagerstätten. Die Diskussion in den drei Fachgruppen entsprechend den gewählten Themenkomplexen wurde zusammengefaßt und reicht von der Definition einer Uranprovinz über die heutige und zukünftige Bedeutung einzelner Provinzen bis zu Empfehlungen zur Erkundungsstrategie und -methodik.

Das Buch repräsentiert in allen Beiträgen den gegenwärtigen Kenntnisstand auf dem Gebiet der Lagerstättenkunde des Urans, ohne einen Anspruch auf Vollständigkeit erheben zu können. Hervorzuheben ist die Fülle von instruktiven geologischen Zeichnungen, die vor allem den lagerstättenbeschreibenden Teil des Buches anschaulich illustrieren.

Wer sich über den modernen Stand der Erforschung von Uranprovinzen der nichtsozialistischen Welt informieren will, findet in diesem Buch, was er sucht.

M. SCHADE

KINDERMANN, F. W. (Hrsg.)

Coal Winning

Proceedings des Informationssymposiums „Neue Methoden und Techniken der Kohlegewinnung in der Europäischen Gemeinschaft" Luxemburg, 23. bis 25. April 1986

Rotterdam—Boston: A. A. Balkema 1987. 250 S., zahlr. Abb., Tab., $ 64.75

In drei Vortragsreihen und drei Rundtischgesprächen wird ein guter Überblick über den Stand und die Entwicklungstendenzen im Steinkohlebergbau der EG vermittelt. Schwerpunkte bilden die Verfahren und Ausrüstungen im unmittelbaren Gewinnungsbereich, die Energieversorgung sowie die Überwachung und Fernsteuerung von Ausrüstungen und schließlich die Arbeiten zum Einrichten und zum Rauben von Streben einschließlich der Probleme des Strebrandes. In den Rundtischgesprächen wurden die Probleme der drei Vortragsreihen jeweils vertieft. Aus den dargestellten umfangreichen technologischen Details und dem Zahlenmaterial sind folgende Schwerpunkte der gegenwärtigen Arbeiten im Steinkohlenbergbau abzuleiten:

- An der Kontrolle und Fernsteuerung der Gewinnungsgeräte wird intensiv gearbeitet. Inzwischen existieren einige Beispiele für die rechnergestützte Überwachung der Gewinnung von über Tage aus und für teilautomatische Arbeitsabläufe.

– Insbesondere im Hinblick auf zukünftige Entwicklungen werden Technologien für die steile Lagerung ausgearbeitet (Frankreich).
– Der ständig steigende Elektroenergiebedarf vor Ort zwingt zu höheren Spannungen. Unter Beachtung auch sicherheitlicher Aspekte ist ein Trend zur 5-kV-Basis erkennbar.
– Zunehmende zu bewegende Massen in den Streben (Ausbaueinheiten) zwingen zu neuen Lösungen beim Einrichten neuer Strebe, bei Raubearbeiten und im Transport.
– Der Übergang Streb/Strecke bildet einen Schwerpunkt von Untersuchungsarbeiten. Neue technische Lösungen für die Strebrandtechnik werden vorgestellt.

Insgesamt bietet dieses Buch einen guten Überblick über den gegenwärtigen Stand der Strebbautechnologie. Die erkennbaren Trendentwicklungen sind auch für den übrigen Untertagebau interessant.

H. GERHARDT

ROV '86

Remotly Operated Vehicles
Technology Requirements — Present and Future

London: Graham & Trotman 1986. 366 S., zahlr. Abb., £ 39.— ($ 76.—)

Die Organisatoren der vom 24.—26. 6. 1986 in Aberdeen an der Schwelle zur Nordsee durchgeführten Konferenz sind Vertreter der führenden westeuropäischen und nordamerikanischen Offshore- und Meerestechnik-Organisationen.
ROV's (unbemannte ferngesteuerte Tauchfahrzeuge) werden seit der zweiten Hälfte der 70er Jahre verstärkt für kommerzielle Aufgaben im Schelfbereich, insbesondere im Offshore-Sektor eingesetzt. Sie werden über Kabel mit Energie versorgt und von einem auf der Versorgungseinheit befindlichen Operator über Telemetrie-Signale ferngesteuert. Entsprechend dem heutigen Stand der Technik gehören zur Standardausrüstung verschiedene rechnergestützte Meßsysteme, Photo- und Fernsehsysteme zur Beobachtung der Umwelt und Manipulatoren zur Ausführung von Wartungs- und Reparaturarbeiten an technischen Anlagen oder zur Entnahme von Wasser- und Bodenproben für Forschungszwecke.

In mehreren Beiträgen zeichnen sich die Tendenzen auf dem Gebiet der ROV-Anwendung ab:

– Verbesserung des Managements und der Einsatztechnologie
– Überwindung von Einsatzgrenzen nicht nur durch Verbesserung der Eigenschaften der ROV's, sondern durch ROV-gerechte Konstruktion der technischen Unterwasser-Anlagen (Schaffung integrierter Systeme)
– Verfeinerung der Meß- und Steuersysteme durch Weiterentwicklung der Sensor- und Rechentechnik
– Entwicklung frei schwimmender und adaptiv gesteuerter ROV's

Das Konferenzmaterial bietet dem Meerestechniker einen interessanten Überblick über den gegenwärtigen Stand und künftige Entwicklungen auf dem ROV-Sektor und seinen Randgebieten, der Unterwasser-Meß- und Steuerungstechnik. Leider sind die Vorträge auf mehr oder weniger aktuelle kommerzielle Anwendungsfälle im Schelfbereich beschränkt. Die Konferenz wird deshalb dem im Untertitel ausgedrückten Anspruch nur bedingt gerecht. Dem erst am Beginn einer sprunghaften Entwicklung stehenden Einsatz von ROV's bei der Untersuchung der aktiven mittelozeanischen Rücken wird nicht die gebührende Beachtung geschenkt.

G. EHRKE

LASER 87

Laser — Optoelektronik — Mikrowellen
8. Internationaler Kongreß und Internationale Fachmesse
22.—26. Juni 1987 in München

Das der Redaktion zur Verfügung gestellte Kongreßmaterial spiegelt die außerordentliche Breite der dort behandelten Themen wider. Es wurden große Bereiche der Industrie, Wissenschaft, Medizin und Nachrichtentechnik vorgestellt. Der Kongreß machte auch deutlich, daß die Lasertechnologie in immer stärkerem Maße in der industriellen Forschung Eingang findet. Das Spektrum reichte von Lichtwellenleitern für optische Informationsübertragungen, interferometrische Meßverfahren als hochgenaue Meßmethode, Anwendung der Laser in der Umweltmeßtechnik (Messungen von Aerosolen bzw. Rauchfahnen usw.), medizinische Problemstellungen bis zu Fragen der Weiterentwicklung sowie des perspektivischen Einsatzes von Laser bis zum Jahre 2000.

B. KÖLBEL

Buchbesprechungen

MILANOVSKIJ, E. E.

Riftogenez v istorii Zemli. Riftogenez v podvižnych pojasach (Riftgenese in der Erdgeschichte. Riftgenese in Mobilzonen)

Moskva: Izdat. Nedra 1987. 297 S., 56 Abb., 3 Tab., 245 Lit.,

GRAČEV, A. F.

Riftovye zony Zemli (Die Riftzonen der Erde)

Moskva: Izdat. Nedra 1987. 2. überarbeitete und ergänzte Auflage. 285 S., 87 Abb., 33 Tab., 207 Lit., Rbl. 3.50 (M 17.50)

Zwei im selben Jahr, im selben Verlag und unter nahezu dem gleichen Titel erschienene Bücher zu vergleichen, ist von besonderem Interesse, zumal dann, wenn sie sich im theoretischen Ansatz, in der Methodik und der regionalen Basis so wesentlich voneinander unterscheiden, wie diese beiden Titel.
Das Buch von MILANOVSKIJ ist Teil einer mehrbändigen Regionalanalyse (Kontinente 1976, alte Tafeln 1983) des Phänomens der Riftgenese, ausgehend von den klassischen Vorstellungen der Geosynklinalentwicklung und methodisch der phänomenologischen Beschreibung struktureller und entwicklungsgeschichtlicher Merkmale folgend. In der großregionalen Gliederung werden die Uralo-Mongolische Mobilzone, Mitteleuropa und der Tethysgürtel einschließlich der angrenzenden Metatafelgebiete behandelt, allerdings mit deutlichen Unterschieden in der Aufbereitung der internationalen Literatur.
Auch für GRAČEV bildet die regional vergleichende Analyse der Riftzonen die Faktengrundlage, nur beschreitet er einen gänzlich anderen Lösungsweg: die Komplexanalyse der für ozeanische und kontinentale Riftsysteme weltweit vorliegenden stofflichen und geophysikalischen Daten. Entsprechend bilden der Vulkanismus als wichtigster Indikator der Paläorifte sowie die Interpretation tiefentektonischer Prozesse für die Riftgenese die Schwerpunkte des Buches. Die geochemische Ableitung einer Evolutionsreihe globaler Riftsysteme und deren genetische Korrelation mit thermischen Prozessen im Kruste—Mantel-Bereich erscheinen dabei besonders produktiv.

Somit sind die Unterschiede zweier demselben Gegenstand verpflichteten Arbeiten kaum größer vorstellbar. Auf der einen Seite eine zuverlässige, doch konservative Analyse mit weitgehend kinematischen Aussagen zum Riftprozeß als das strukturelle Ergebnis des räumlich-zeitlichen periodischen Wechsels von Ausweitung und Einengung. Auf der anderen Seite die Erarbeitung eines geodynamischen Modells der Riftgenese auf geochemisch-geophysikalischer Grundlage. MILANOVSKIJ fragt nach dem Wo und Wann, GRAČEV nach dem endogenen Wie. Für MILANOVSKIJ ist der Geosynklinalprozeß bestimmend für die Krustenentstehung, GRAČEV dagegen fordert dazu auf „sich von der Geosynklinaltheorie zu trennen".
Trotz dieser so grundverschiedenen Sichtweisen stellen jedoch beide Bücher — jedes auf seine Weise — einen verdienstvollen Beitrag zur Lösung des für die moderne Geotektonik so wichtigen Problems der Riftgenese dar.

H.-J. TESCHKE

WEZEL, F.-C. (Hrsg.)

The Origin of Arcs

Amsterdam—Oxford—New York—Tokyo: Elsevier Science Publ. 1986. XX + 567 S., Dfl. 180.— ($ 72.—)

(Developments in Geotectonics 21)

Dieser Band enthält 23 Beiträge eines 1986 in Urbino (Italien) veranstalteten internationalen Symposiums über Insel- und Gebirgsbögen. Sie sind fünf thematischen Gruppen zugeordnet: Geotektonische Konzepte, Geophysik und Geochemie, Alpin-Mediterranes Gebiet, Karibischer Raum, Pazifik. Ziel des Symposiums war es u. a., Antworten auf folgende Fragen zu finden: Sind es primäre oder induzierte Erscheinungen? Welche Beziehungen bestehen zwischen Insel- und Gebirgsbögen? Welche Hauptdeformationskraft hat zu ihrer Bildung geführt? Was sind Benioff-Zonen wirklich und welche tektonische Bedeutung besitzen sie? Dazu werden vor allem im ersten Teil unterschiedliche, von den kon-

ventionellen Vorstellungen teilweise beträchtlich abweichende Auffassungen geäußert: Orogenese als diapirischer Prozeß auf der Grundlage der Expansionshypothese (Krikogenese); extreme Langlebigkeit bogenförmig-orogener Hebungszonen; Orogen-Stränge als Kompressions- und Scherzonen der Lithosphäre zwischen zwei geotektonischen Großeinheiten (Pazifik und Gondwana) über zwei riesigen Konvektionszellen im Mantel, Einfluß der Erdrotation und von Rotationsänderungen auf geotektonische Vorgänge in den Subduktionszonen, Vulkanismus, Seismizität, Neotektonik) u. a.
Im zweiten Teil werden Lithosphärenmächtigkeiten, Dynamik und Streß an konvergierenden Plattengrenzen, Paläomagnetismus sowie geochemische und isotopische Änderungen in Vulkaniten der Inselbögen für die Beantwortung der o. g. Fragen herangezogen. Die regional eingeordneten Beiträge behandeln sehr ausführlich den alpin-mediterranen Raum. Zwei Beiträge über die Karibik beleuchten die Bedeutung der Sedimentbeschaffenheit für den Charakter der Deformationen bzw. die relativ komplizierten Bewegungen der Karibischen Platte. Von den den Pazifik betreffenden Beiträgen verdient der über dessen Genese in Verbindung mit einer Erdexpansion besonderes Interesse. Im übrigen werden neue Fakten sowie offene Fragen zu den Tiefseegräben und Inselbögen dargelegt.
Dieser neue Band der „Developments in Geotectonics" gibt nicht nur Auskunft über die Vielfalt bogenförmiger geotektonischer Erscheinungen, sondern beleuchtet zugleich auch die unterschiedlichen Möglichkeiten ihrer genetischen Interpretation.

G. KATZUNG

PUŠČAROVSKIJ, JU. M. (Red.)

Zakonomernosti formirovanija struktury kontinentov v neogee (Gesetzmäßigkeiten der Bildung von Strukturen in den Kontinenten der Neogäa)

Moskva: Izdat. Nauka 1986. 204 S., Rbl. 3.30

In dem Sammelband mit 18 Arbeiten kompetenter Geowissenschaftler der Akademie der Wissenschaften der UdSSR werden auf der Basis neuester Daten Prozesse der stofflich-strukturellen Formung der kontinentalen Kruste im Zeitraum ab oberem Proterozoikum analysiert und die Hauptgesetzmäßigkeiten ihrer Entwicklung diskutiert.
Nach der Darstellung allgemeiner Gesetzmäßigkeiten (PEJVE, KNIPPER u. a.; BELOUSOV) werden hauptsächlich der Bau und die Entwicklung von Faltengürteln im Territorium der UdSSR und angrenzender Räume charakterisiert. Schwerpunktgebiete sind die paläozoische Formung der Uralo-Mongolischen Zone (DOBRECOV; KARAULOV; ABDULIN & ZAJCEV; DERGUNOV u. a.; SAMYGIN & LEJTEC), die paläozoische (ADAMIJA, BELOV u. a.) und mesokänozoische Entwicklung (GAMKRELIDZE; ŠOLPO) der mediterranen Faltenzone und der mobilen Krustenteile in NE-Sibirien (TIL'MAN) sowie der randkontinentalen Teile am Pazifik (ZINKEVIČ, MAZAROVIČ u. a.). Die regionalen und methodischen Untersuchungen erfassen die Bildung und das endogene Regime der kontinentalen Kruste, die Haupttypen der Strukturen (GARECKIJ & ŠLESINGER), die Rolle von Tiefenbrüchen und Ophioliten, Deformationsmechanismen (BOGDANOV) mit der Anlage und Entwicklung mobiler Gürtel (KROPOTKIN; BOŽKO) und Probleme des Zusammenhanges von Tektonik und Metallogenie (RUNDKVIST, MARKOV u. a.; TVALČRELIDZE & ŠČEGLOV).
Der Band gibt die gesamte Breite der z. Z. in der UdSSR vertretenen Theorien, Modelle usw. zur geotektonischen Entwicklung der kontinentalen Kruste wieder. Die Arbeiten sind gestrafft und gut lesbar.

M. KRAUSS

Autorenkollektiv

Grundzüge der geotektonischen Entwicklung Mitteleuropas
Ausgewählte Minerallagerstätten der Deutschen Demokratischen Republik

Wissenschaftlich-Technischer Informationsdienst, Berlin, Reihe A, 28 (1987) 3, 55 S.

Im Jahre 1986 erschien der Band 5 „Central Europe" der durch das Institute of Mining and Metallurgy unter der Redaktion von F. W. DUNNING & A. M. EVANS herausgegebenen „Mineral Deposits of Europe". Ziemlich umfangreich und daher nicht gerade billig, wird dieser Band verständlicherweise nicht in allen geowissenschaftlichen Bibliotheken unseres Landes verfügbar sein. Es ist daher zu begrüßen, daß die Redaktion des Wissenschaftlich-Technischen Informationsdienstes des ZGI die Möglichkeit genutzt hat, den von K. SCHMIDT, B. KÖLBEL, L. BAUMANN, M. KRAFT, J. RENTZSCH & S. LÄCHELT verfaßten DDR-Beitrag zu diesem Werk in deutscher Sprache zu publizieren.
In gestraffter Form wird im Kap. 1 ein Überblick über die geotektonische Entwicklung und Gliederung Mitteleuropas sowie die für diesen Raum typischen geologisch-minerogenetischen Merkmale der einzelnen tektonischen Stockwerke vom Präkambrium bis zu alpidischer Zeit gegeben. Kap. 2 behandelt DDR-spezifische Probleme: zum einen wird über die geotektonisch-minerogenetische Gliederung des Gesamtterritoriums, die speziellen minerogenetischen Entwicklungsstadien sowie die auf dieser Grundlage durchgeführte Ausscheidung und Abgrenzung regionaler minerogenetischer Einheiten informiert, zum anderen werden am Beispiel ausgewählter Minerallagerstätten die für die DDR charakteristischen prävariszischen, variszischen und postvariszischen Mineralisationen beschrieben. Abbildungen und Tabellen illustrieren den Text und tragen damit zum Verständnis der Thematik zusätzlich bei. Nicht nur Studenten werden dieses Heft begrüßen, auch dem gestandenen Geologen bietet es auf Grund der wissenschaftlichen Breite und Aktualität des Stoffes Gelegenheit, Bekanntes aufzufrischen und Neues kennenzulernen.
Der Vertrieb erfolgt ausschließlich durch das Zentrale Geologische Institut.

D. FRANKE

EDWARDS, R.; ATKINSON, K.

Ore Deposit Geology and its Influence on Mineral Exploration

London—New York: Chapman & Hall Ltd. 1986. 466 S., zahlr. Abb., £ 19.50

Der Titel des Buches weist bereits darauf hin, daß es sich nicht um eine erneute „Erzlagerstättenkunde" handelt, sondern daß hier Belange der Geologen im Vordergrund stehen, die in der Lagerstättensuche tätig sind oder werden sollen. So gilt die Einführung der Erläuterung der Beziehungen zwischen lagerstättengenetischen und -kontrollierenden Faktoren, ökonomischen, bergbaulichen und metallurgischen Gesichtspunkten, die bei der Suche im weitesten Sinne berücksichtigt werden müssen. In den folgenden neun Kapiteln sind die Haupttypen der Erzlagerstätten dargestellt, wobei jeweils von den modernsten genetischen Vorstellungen ausgegangen, an zahlreichen Beispielen die konkreten Verhältnisse erläutert (darunter Paragenesen, Alteration, Tektonik) und schließlich Hinweise auf die Nutzung verschiedener Methoden für die Suche gegeben werden (z. B. Phototektonik, Geochemie, Geophysik u. a.), teilweise mit Ansätzen für die Lagerstättenbewertung. Dabei wird bewußt auf eine nicht zu erreichende Vollständigkeit verzichtet. Zahlreiche Risse, Schnitte, Diagramme, Schemata und Photos ergänzen den knappen, inhaltsreichen Text hervorragend.
Im Kapitel „magmatische Lagerstätten" werden Chromit-, Nickelsulfid-Lagerstätten und Kimberlite, unter „magmatisch-hydrothermalen" Lagerstätten Porphyr-Kupfer-, Porphyr-Molybdänsowie vulkanogene Sulfid-Lagerstätten dargestellt. Bei den hydrothermalen Ganglagerstätten werden ausschließlich Gold-Mineralisationen beschrieben. Es folgen Seifenlagerstätten, an Sedimente gebundene Cu- und Pb—Zn-Mineralisationen (Kupferschiefer, Mississippi-Valley-Typ), Verwitterungslagerstätten (Bauxit, lateritische Ni-Lager, Kaolin), sedimentäre Eisenerze (mit Schwerpunkt auf gebänderten proterozoisch-paläozoischen Erzen), an Sedimente gebundene U-Mineralisationen sowie Skarnlagerstätten. Das letzte Kapitel gilt der Planung und Durchführung von Such-/Erkundungsprogrammen mit Darstellung von Etappen, Techniken, Kosten u. a. Jedem Kapitel ist ein Literaturverzeichnis angefügt. Insgesamt ein sehr anregendes, modernes Erzlagerstättenbuch, auch wenn ausschließlich Lagerstätten der westlichen Hemisphäre betrachtet werden.

G. RÖLLIG

...

The Crust — the Significance of Granite Gneisses in the Lithosphere

Athen: Theophrastus Publ. S. A. 1985. 714 S.

Der 1. Abschnitt des vorliegenden Bandes enthält sieben Beiträge zur Granitgenese als einem zentralen Problem der Lithosphärenentwicklung. Gegenstand vieler dieser Artikel ist zunächst eine geologisch-petrographische bzw. petro-geochemische Charakteristik sowohl archaisch-proterozoischer Granit-Gneise und Anatexite (Brasilianischer, Indischer, Nordamerikanischer Schild) als auch herzynischer Granitoide unterschiedlicher Genese (Französisches Zentralmassiv, Pyrenäen, SW-Spanien). Auf dieser Grundlage werden petrogenetische Schlußfolgerungen gezogen und mögliche Genesemodelle der Lithosphäre aufgezeigt.
Auch solche petrogenetischen Spezialfragen wie die Deutung der in Granitoiden des Französischen Zentralmassivs auftretenden Cordierit-Quarz-Knauern als metasomatische Bildungen (ähnlich den Turmalinsonnen in anderen Massiven) (DIDIER, DUPRAZ) oder die Betrachtung über vermutlich im Resultat von herzynen Granitisierungsprozessen entstandenen ovalen Oberflächenstrukturen von silifizierten Nebengesteinspaketen im rezenten Erosionsanschnitt (SANCHEZ-CELA) werden behandelt. Die 16 Beiträge des 2. Abschnittes stammen ausschließlich von chinesischen Autoren und befassen sich in erster Linie mit mesozoischen Granitoidkomplexen

im Südosten der VR China. Mit Hilfe von geologischen, petrographischen und mineralogischen Merkmalen sowie petrochemischen, geochemischen und isotopengeochemischen Parametern werden verschiedene genetische Typen unterschieden. Grundlage für die Differenzierung ist die Klassifikation nach CHAPPEL & WHITE (I-, S-Typen) wie auch die nach XU KE CHIN (kontinentaler Krustentransformations-Typ, Übergangskruste und Manteltyp).
In mehreren Arbeiten wird die besondere Bedeutung der Gehalte an Seltenen Erden und abgeleiteter Koeffizienten bei der Klärung genetischer Fragen aufgezeigt (z. B. JILI JIAYUAN).
Die 13 Beiträge des 3. Abschnittes beschäftigen sich mit der metallogenetischen Bedeutung chinesischer Granitoidkomplexe im allgemeinen und der Mineralisation der Elemente W, Ta—Nb, Mo, Sn, Au im besonderen. In speziellen Artikeln wird auf pegmatitische Mineralisation, Be-Gehalte in altersverschiedenen Granitoiden und den ersten chinesischen Fund von gediegen Aluminium eingegangen.

H. MÜLLER

...

Untersuchung des gestörten Lockergesteins

Geophys. u. Geol., Berlin, Bd. III (1987) 4, 236 S., M 48.—

Dieser Band enthält fünf längere Originalbeiträge zur Thematik Lagerungsverhältnisse und Lagerungsstörungen im känozoischen Deckgebirge und geht auf ein Vortragsprogramm von geowissenschaftlichen Weiterbildungslehrgängen an der Karl-Marx-Universität Leipzig zurück. Diese Publikation ist sehr zu begrüßen, ist doch das Interesse an diesen Fragen groß und überaus weit gestreut.
Ein umfangreicher Beitrag von L. EISSMANN (70 S.) leitet das Heft mit einer geologischen Analyse der exogenen und endogenen Lagerungsstörungen im Lockergebirge ein. Die vorhandenen Formen werden klassifiziert, alle ausgehaltenen Typen anhand meist mehrerer Beispiele erläutert und durch gut ausgewählte Abbildungen illustriert. Das bekannte didaktische Geschick des Autors wird hier sichtbar. Ein umfangreiches Literaturverzeichnis von rd. 130 Zitaten erleichtert dem Leser ein weiteres Eindringen in die Problematik.
Die Beiträge von H. RISCHE (Seismische Verfahren, 50 S.), K. SCHÖSSLER (Gravimetrische Verfahren, 29 S.) sowie von F. JAKOBS, H. PETZOLD & K.-H. NITSCH (Geoelektrische Verfahren, 35 S.) informieren gut über die wachsende Bedeutung von Verfahren der angewandten Geophysik bei der Erkundung von Lagerungsverhältnissen und -störungen im Lockergebirge. Für den Nicht-Geophysiker verständlich, werden Wirkprinzipien, methodische und technische Grundlagen sowie Ergebnisse diskutiert und an gut ausgesuchten Interpretationsbeispielen erläutert.
Der letzte Artikel, ein Beitrag von K.-D. KRESSER, H. MATTERSTEIG, G. HÖNEMANN & G. KUHNE (29 S.) greift ein Thema auf, mit dem Fachleute in der Erkundungspraxis seit langem konfrontiert sind, nämlich, auf welche Weise sich aus Bohrlochmeßkurven Angaben zu den Lagerungsverhältnissen, zu Lagerungsstörungen und zum wahrscheinlichen Störungsmodell ableiten lassen. Die hierzu notwendigen Interpretationsschritte werden erläutert und durch zahlreiche Abbildungen gestützt. Eine gute und seit langem notwendige Darstellung.
Vorliegendes Heft 4 ist für alle im Lockergebirgsstockwerk arbeitenden Geowissenschaftler von großem Interesse. Eine besondere Empfehlung ist daher überflüssig.

H. AHRENS

TER-MARTIROSJAN, Z. G.

Prognozy mechaničeskich processov v massivach mnogofaznych (Prognose mechanischer Prozesse in mehrphasigen Gesteinsmassiven)

Moskva: Izdat. Nedra 1986. 289 S., 80 Abb., 44 Lit., Rbl. 1.30

Das Buch bietet einen guten Überblick über das bodenmechanische Verhalten mehrphasiger Tongesteine. Im Vordergrund steht die mathematische Behandlung der Eigenschaften und Prozesse in derartigen Gesteinen (Teil I: Grundlagen der Mechanik mehrphasiger Böden und Teil II: Prognose der geomechanischen Prozesse in Massiven mit mehrphasigen Böden). Im ersten Teil wird nach einer allgemein physiko-mechanischen Charakteristik speziell das Spannungs-Deformationsverhalten dieser Gesteine mathematisch beschrieben und im zweiten Teil auf Rutschungs- und Böschungsprobleme unter Berücksichtigung der Standdauer und äußeren Belastung der Böschungen angewendet.
In den Abschnitten 9 bis 11 sind speziell Hebungs- und Senkungsvorgänge in derartigen Gesteinen im Zusammenhang mit Grundwasserabsenkung oder -wiederanstieg, Gründungsarbeiten und deren mathematische Fassung behandelt.
Der Text ist methodisch gut aufgebaut, ein Verzeichnis der verwendeten Kennwerte und Einheiten könnte das Einarbeiten in die mathematischen Ableitungen erleichtern. Das Anliegen des Buches besteht darin, das Interesse des geotechnisch orientierten Lesers für die mechanischen Begründungen des Verhaltens mehrphasiger Tongesteine und ihrer mathematischen Fassung zu wecken. Es ist deshalb neben dem Geotechniker, Bergbautechnologen und Bauingenieur auch dem Studierenden dieser Fachrichtungen zu empfehlen.

H. MOLEK

SCHMINCKE, H.-U.
Vulkanismus

Darmstadt: Wissenschaftliche Buchgesellschaft 1986. 164 S., 127 Abb., 6 Tab., 288 Lit., DM 39.—

Der Autor, Professor für Petrologie und Geologie an der Ruhruniversität Bochum und z. Z. Generalsekretär der International Association of Volcanology and Chemistry of Earth's Interior (IAVCEI), vermittelt interessante Einblicke in verschiedene aktuelle Probleme der Vulkanologie. Dargestellt werden neben den allgemeinen Grundlagen vorwiegend geologische und petrologische Aspekte. Das Buch ist eine ausgezeichnete Einführung in die Vulkanologie; es dokumentiert den erreichten Kenntnisstand über die Probleme des Vulkanismus durch eine ausführliche Diskussion der geophysikalisch-stofflichen-strukturellen Ursachen dieser Erscheinungen. Dieses Anliegen wird unterstützt durch instruktive Schemata bzw. Abbildungen, durch Farbphotos ausgezeichneter Qualität, sowie durch ein ausführliches Literaturverzeichnis. Der Autor versteht es, die sehr komplexen geowissenschaftlichen Phänomene, die den Vulkanismus kennzeichnen, anschaulich zu erläutern und auch die Grenzen unseres derzeitigen Wissens aufzuzeigen. Das kommt besonders im Kapitel „Mensch und Vulkan" zum Ausdruck, worin Prognose (forecast) und Vorhersage (prediction) behandelt werden. Derzeit sind Vorhersagen nur mit mittleren Wahrscheinlichkeiten möglich. Der Autor demonstriert dies an den Ausbruch-Vorhersagen für die Vulkane Soufrière (1976) und Mauna Loa (1976), die nicht zutrafen. Grundvoraussetzung jeglicher Prognosen sind demnach „gut instrumentierte" Vulkane zur Ableitung von „Verhaltensdiagrammen", Bedingungen, die derzeit erst wenige Länder realisieren konnten. Abschließend vermittelt der Autor einen Überblick über die mögliche Nutzung der Wärme aus der Erde und behandelt die Grundlagen der erst kürzlich entdeckten Bildung rezenter hydrothermaler Lagerstätten im Pazifik (black smoker) durch Wechselwirkung zwischen Magmawärme einerseits und externem Wasser andererseits.
Das nicht zuletzt auch sehr schöne Buch ist allen Geowissenschaftlern zu empfehlen.

P. KÜHN

RONA, P.

Gidrothermal'naja mineralizacija oblastej spredinga v okeane (Hydrothermale Mineralisation in ozeanischen „Spreading"-Bereichen)

Moskva: Izdat. Mir 1986. 160 S., 14 Abb., 6 Tab., 512 Lit., Rbl. 1.60

Die russische Übersetzung des 1984 bei Elsevier Science Publishers B. V. erschienenen Buches „Hydrothermal Mineralization at Seafloor Spreading Centers" befaßt sich mit dem hochaktuellen Problem des Zusammenhanges submariner hydrothermaler Prozesse und dem sogenannten „Seafloor Spreading". Dieses Buch des bekannten amerikanischen Ozeanologen P. RONA stellt eine gelungene Synthese der in jüngster Vergangenheit in großem Umfang erschienenen Literatur zum Thema der hydrothermalen Erzbildungsprozesse dar, wovon nicht zuletzt das recht umfangreiche Literaturverzeichnis zeugt, das durch den sowjetischen Übersetzer durch eine Reihe z. T. neuester Arbeiten sowjetischer Wissenschaftler vervollständigt wurde. Anhand von geophysikalischen und tektonischen Daten, aber auch unter Verwendung von Erkenntnissen laborativer Untersuchungen versucht der Autor den eigentlichen Mechanismus dieser Erzbildungsprozesse zu klären sowie die Fakten herauszuarbeiten, die auf die Zusammensetzung und Verteilung hydrothermaler Vererzungen einwirken. Als entscheidender Faktor wurde vom Vf. die Geschwindigkeit des „Spreading" erkannt. Das ermöglichte von bis dahin lediglich zufälligen Funden zur gezielten Suche von Hydrothermenaustritten auf dem Meeresgrund überzugehen.
Die Untersuchungen von RONA stellen somit ein Beispiel der erfolgreichen Anwendung der Theorie der Plattentektonik für die Suche von Lagerstätten auf dem Meeresboden dar.
Da die Aussagen des Autors lediglich auf der Untersuchung von 55 verschiedenen Fundpunkten beruhen, wobei das entnommene Material im wesentlichen von der Oberfläche des Meeresbodens stammt, sind einige Aussagen des Autors, wie er selbst bemerkt, noch nicht völlig bewiesen und bedürfen einer Bestätigung durch weitere Untersuchungen. Dieses Buch ist von sehr großem Interesse für jeden Geologen, Geochemiker, Metallogenetiker und Geophysiker.

W. SCHULZE

Bei Fragen zur Produktsicherheit wenden Sie sich bitte an:
If you have any questions regarding product safety,
please contact:

Walter de Gruyter GmbH
Genthiner Straße 13
10785 Berlin
productsafety@degruyterbrill.com